Surface Water Quality Modelling

Surface Water Quality Modelling

Editor

Karl-Erich Lindenschmidt

MDPI • Basel • Beijing • Wuhan • Barcelona • Belgrade • Manchester • Tokyo • Cluj • Tianjin

Editor
Karl-Erich Lindenschmidt
Global Institute for Water
Security
School of Environment and
Sustainability
University of Saskatchewan
Saskatoon
Canada

Editorial Office
MDPI
St. Alban-Anlage 66
4052 Basel, Switzerland

This is a reprint of articles from the Special Issue published online in the open access journal *Water* (ISSN 2073-4441) (available at: www.mdpi.com/journal/water/special_issues/surface_water).

For citation purposes, cite each article independently as indicated on the article page online and as indicated below:

LastName, A.A.; LastName, B.B.; LastName, C.C. Article Title. *Journal Name* **Year**, *Volume Number*, Page Range.

ISBN 978-3-0365-6947-5 (Hbk)
ISBN 978-3-0365-6946-8 (PDF)

Cover image courtesy of Karl-Erich Lindenschmidt

Contents

About the Editor

Karl-Erich Lindenschmidt

Karl-Erich Lindenschmidt is a professor at the University of Saskatchewan. His main research areas are (i) river ice monitoring and modelling and (ii) surface water quality modelling. He holds a Bachelor of Science in Mechanical Engineering from the University of Manitoba. He went on to take a Master of Applied Science in Mechanical Engineering from the University of Toronto and a Ph.D. in Environmental Engineering from the Technical University of Berlin, Germany.

Karl worked in Europe for 16 years at several consultancies and research facilities. Before his appointment at the University of Saskatchewan in 2012, Karl was with Manitoba Water Stewardship for three years as a Hydrologic Modelling Research Engineer where he developed river ice monitoring and modelling capabilities for the Red and Dauphin rivers. He is currently developing ice-jam flood hazard and risk methodologies for many rivers across Canada and Alaska.

Preface to "Surface Water Quality Modelling"

This Special Issue showcases some advanced methodologies used in surface water quality modelling. Its aim is to introduce several topics that are new in the field of surface water quality modelling and help spur new developments in each of the topics. It is intended for a readership consisting of researchers, academicians, consultants and government agency representatives. The editor of this Special Issue wishes to thank all authors, reviewers and academic editors in their contributions, without whom the publishing of this collective work would not have come to pass.

Karl-Erich Lindenschmidt
Editor

water

Editorial

Surface Water Quality Modelling

Karl-Erich Lindenschmidt

Global Institute for Water Security, School of Environment and Sustainability, University of Saskatchewan, 11 Innovation Boulevard, Saskatoon, SK S7N 3H5, Canada; karl-erich.lindenschmidt@usask.ca

Surface water quality modelling has become an important means of better understanding aquatic and riparian ecosystem processes at all scales, from the micro-scale (e.g., bottom sediment dynamics), to the meso-scale (e.g., algal bloom growth) and the macro-scale (e.g., the role of cascading reservoirs in sediment transport). Increasingly, surface water quality models are being coupled to other models (e.g., hydrological models) to determine catchment area impacts on water quality. These impacts include future climate change and land-use developments. Coupling with water resource dynamics models also provides insight into changes in water supply and demand and flow regulation as they relate to surface water quality. Modelling the quality of surface waters under ice-covered conditions has also gained special attention due to the increased realization that a holistic all-year perspective is required to deepen our understanding of aquatic ecosystem functioning (e.g., the impact of lake ice phenology on spring succession of phytoplankton) (excerpts and adaptations from my Global Water Futures https://gwf.usask.ca/ (accessed on 16 February 2023) reporting and https://www.mdpi.com/journal/water/special_issues/surface_water (accessed on 16 February 2023) announcement).

Table 1 summarises the contents of the papers [1–10] submitted to this Special Issue into categories of global region, type of water body and model or modelling approach used. From North America, modelling exercises on rivers in Canada have been reported, particularly the Athabasca [9] and South Saskatchewan [1] rivers in western Canada. The Athabasca River is a natural flowing river draining a catchment area that is predominantly forested, whereas the South Saskatchewan River is a prairie river mostly draining agricultural lands and regulated by the Gardiner Dam, which impounds Lake Diefenbaker. Both studies look at the transport and fate of heavy metals (copper, nickel and vanadium) to verify the sources of these metals, with copper stemming mainly from anthropogenic and nickel and vanadium from geogenic sources.

The water quality of a Canadian reservoir, Buffalo Pound Lake, was also modelled [7]. Buffalo Pound Lake receives water from Lake Diefenbaker (mentioned above) and is an important drinking water source for southern Saskatchewan. Managing water transfers from Lake Diefenbaker is an important aspect when modelling the algal-nutrient dynamics in Buffalo Pound Lake.

China is represented in this Special Issue by the modelling applications of surface waters in Guangzhou Province [2] of southern China and the estuary Sanmen Bay in Zhejiang Province of eastern China [3]. Both applications use chemical oxygen demand (COD) as a determinant to assess the impact of urban pollutants on aquaculture facilities. A study of the transport of nutrients, particularly ammonium nitrogen, in overland runoff from the Haihe River basin near Beijing and the Poyang Lake basin, which drains into the Yangtze River in east central China [8], is also available. The Haihe River basin study looks more at reducing diffuse loadings, whereas the Poyang Lake basin study investigated point loadings as mitigation options to control sediment and nutrient loadings in receiving waters.

Citation: Lindenschmidt, K.-E. Surface Water Quality Modelling. *Water* **2023**, *15*, 828. https://doi.org/10.3390/w15040828

Received: 13 February 2023
Accepted: 15 February 2023
Published: 20 February 2023

Table 1. Summary of the models used, surface waters investigated and study regions of the articles in this Special Issue, *Surface Water Quality Modelling*.

Model	Region						
	Canada	**China**	**Vietnam**	**India**	**South Korea**	**Malawi**	**Nordic Europe**
HEC-RAS	river [1]						
WASP	river [1]			river [5]	river [5]		
WASP (Quasi-2D)	river [9]						
statistical		river [2]					
Delft 3D		estuary [3]					
Machine Learning			river [4]				
conceptual						lake [6]	
CE-QUAL-W2	reservoir [7]						
SPARROW		overland [8]					
various							river [10]

Other river water-quality modelling studies reported from Asia include the Yamuna River in northern India [5], the Ara River in South Korea [5] and the La Buong River in Vietnam [4], listed in the order of wetness ranging, respectively, from dry, moderately wet to wet. Typical water-quality constituents such as biological oxygen demand (BOD) and dissolved oxygen (DO) with different arrays of total or dissolved nutrients were modelled for different seasons (dry/wet, spring/summer/autumn/winter).

One study from Africa (Lake Malombe in Malawi) [6] and another from Nordic Europe (various rivers in Norway, Sweden, Finland, eastern Germany and Poland) [10] are included in this Special Issue. The first study links overarching demographic, socio-economic, climatic and other factors to the lake's ecosystem health. The second study focuses more on river ice processes and reveals the importance of increased water temperatures from urban heat islands and increased salinity due to fertilizer applications in agricultural areas for long-term trends in water parameters.

The Special Issue also provides a spectrum of different model applications, from the more conceptual [6], to empirical approaches [2,4,8] to the more deterministic methodologies [1,3,5,7,9]:

Conceptual modelling:

The conceptual model uses driver, pressure, state, impact, and response (DPSIR) indicators to assess the ecosystem health of Lake Malombe in Malawi [6].

Empirical modelling:

A diagnostic model consisting of visible near-infrared spectroscopy combined with partial least squares discriminant analysis was applied to determine the water quality status of 127 surface water sites [2]. Twelve machine-learning approaches were employed to assess the water quality of the La Buong River in Vietnam [4]. SPARROW, too, is an empirical model relating catchment area characteristics and discharges to nutrient loading in large river basins [8].

Deterministic modelling:

The WASP (Water Quality Analysis Simulation Program), originally developed by the US Environmental Protection Agency, is a popular model applied to rivers and lakes to track water quality changes in surface water bodies [1]. Its modules of HEAT, EUTRO and TOXI allow a diverse set of processes and constituents to be modelled. The discretisation scheme also allows different dimensionality schemes, from one-, two- and three-dimensional schemes but also a quasi-two-dimensional scheme, where one-dimensional equations are applied across a two-dimensional discretisation network providing flexibility in model setup at lower computational expenditure [9]. Coupling to a hydraulic model such as HEC-RAS gives the WASP modelling system a better description of the hydraulic flow fields [1]. The two-dimensional CE-QUAL-2D and three-dimensional Delft 3D models provide more descriptive power by including changing constituent concentrations in depth, such as in lakes and reservoirs [7], plus lateral gradients, such as in estuaries [3], not just longitudinally, which is often the only dimension requirement for modelling rivers [1].

Acknowledgments: The Guest Editor wishes to express their gratitude to all authors' and reviewers' contributions to this Special Issue.

Conflicts of Interest: The author declares no conflict of interest.

References

1. Prajapati, S.; Sabokruhie, P.; Brinkmann, M.; Lindenschmidt, K.-E. Modelling Transport and Fate of Copper and Nickel across the South Saskatchewan River Using WASP—TOXI. *Water* **2023**, *15*, 265. [CrossRef]
2. Han, X.; Chen, X.; Ma, J.; Chen, J.; Xie, B.; Yin, W.; Yang, Y.; Jia, W.; Xie, D.; Huang, F. Discrimination of Chemical Oxygen Demand Pollution in Surface Water Based on Visible Near-Infrared Spectroscopy. *Water* **2022**, *14*, 3003. [CrossRef]
3. Yao, Y.; Zhu, J.; Li, L.; Wang, J.; Yuan, J. Marine Environmental Capacity in Sanmen Bay, China. *Water* **2022**, *14*, 2083. [CrossRef]
4. Khoi, D.N.; Quan, N.T.; Linh, D.Q.; Nhi, P.T.T.; Thuy, N.T.D. Using Machine Learning Models for Predicting the Water Quality Index in the La Buong River, Vietnam. *Water* **2022**, *14*, 1552. [CrossRef]
5. Iqbal, M.M.; Li, L.; Hussain, S.; Lee, J.L.; Mumtaz, F.; Elbeltagi, A.; Waqas, M.S.; Dilawar, A. Analysis of Seasonal Variations in Surface Water Quality over Wet and Dry Regions. *Water* **2022**, *14*, 1058. [CrossRef]
6. Kosamu, I.B.M.; Makwinja, R.; Kaonga, C.C.; Mengistou, S.; Kaunda, E.; Alamirew, T.; Njaya, F. Application of DPSIR and Tobit Models in Assessing Freshwater Ecosystems: The Case of Lake Malombe, Malawi. *Water* **2022**, *14*, 619. [CrossRef]
7. Terry, J.; Davies, J.-M.; Lindenschmidt, K.-E. Buffalo Pound Lake—Modelling Water Resource Management Scenarios of a Large Multi-Purpose Prairie Reservoir. *Water* **2022**, *14*, 584. [CrossRef]
8. Yin, J.; Chen, H.; Wang, Y.; Guo, L.; Li, G.; Wang, P. Ammonium Nitrogen Streamflow Transport Modelling and Spatial Analysis in Two Chinese Basins. *Water* **2022**, *14*, 209. [CrossRef]
9. Sabokruhie, P.; Akomeah, E.; Rosner, T.; Lindenschmidt, K.-E. Proof-of-Concept of a Quasi-2D Water-Quality Modelling Approach to Simulate Transverse Mixing in Rivers. *Water* **2021**, *13*, 3071. [CrossRef]
10. Lindenschmidt, K.-E.; Alfredsen, K.; Carstensen, D.; Choryński, A.; Gustafsson, D.; Halicki, M.; Hentschel, B.; Karjalainen, N.; Kögel, M.; Kolerski, T.; et al. Assessing and Mitigating Ice-Jam Flood Hazards and Risks: A European Perspective. *Water* **2023**, *15*, 76. [CrossRef]

water

MDPI

Article

Modelling Transport and Fate of Copper and Nickel across the South Saskatchewan River Using WASP—TOXI

Saurabh Prajapati [1,2], Pouya Sabokruhie [2], Markus Brinkmann [1,2,3,4,*] and Karl-Erich Lindenschmidt [1,2]

1 School of Environment and Sustainability, University of Saskatchewan, Saskatoon, SK S7N 0X1, Canada
2 Global Institute for Water Security, University of Saskatchewan, Saskatoon, SK S7N 3H5, Canada
3 Toxicology Centre, University of Saskatchewan, Saskatoon, SK S7N 5B3, Canada
4 Centre for Hydrology, University of Saskatchewan, Saskatoon, SK S7N 1K2, Canada
* Correspondence: markus.brinkmann@usask.ca

Abstract: The South Saskatchewan River (SSR) is one of the most important river systems in Saskatchewan and, arguably, in Canada. Most of the Saskatchewan residents, industries, and power-plants depend on the SSR for their water requirements. An established 1D modelling approach was chosen and coupled with the Hydrologic Engineering Center's River Analysis System (HEC-RAS). The WASP (Water Quality Analysis Simulation Program) stream transport module, TOXI, is coupled with flow routing for free-flow streams, ponded segments, and backwater reaches and is capable of calculating the flow of water, sediment, and dissolved constituents across branched and ponded segments. Copper and nickel were chosen as two metals with predominantly anthropogenic (agriculture, mining, and municipal and industrial waste management) and geogenic (natural weathering and erosion) sources, respectively. Analysis was carried out at ten different sites along the South Saskatchewan River, both upstream and downstream of the City of Saskatoon, in the years 2020 and 2021. Model performance was evaluated by comparing model predictions with concentrations of copper and nickel measured in a previously published study. The model performed well in estimating the concentrations of copper and nickel in water samples and worked reasonably well for sediment samples. The model underestimated the concentration values at certain segments in both water and sediment samples. In order to calibrate the model more accurately, extra diffusive contaminant loads were added. While several default parameter values had to be used due to the unavailability of primary historical data, our study demonstrates the predictive power of combining WASP—TOXI and HEC-RAS models for the prediction of contaminant loading. Future studies, including those on the impacts of global climate change on water quality on the Canadian prairies, will benefit from this proof-of-concept study.

Keywords: copper; nickel; trace metals; pesticides; WASP; HEC-RAS

Citation: Prajapati, S.; Sabokruhie, P.; Brinkmann, M.; Lindenschmidt, K.-E. Modelling Transport and Fate of Copper and Nickel across the South Saskatchewan River Using WASP—TOXI. *Water* **2023**, *15*, 265. https://doi.org/10.3390/w15020265

Academic Editors: Bommanna Krishnappan and Andrea G. Capodaglio

Received: 30 November 2022
Revised: 3 January 2023
Accepted: 5 January 2023
Published: 8 January 2023

1. Introduction

Contamination of the environment with various trace metals has been of significant concern for many decades due to their potential to cause deleterious impacts on exposed wildlife and humans. Trace metals have several unique chemical properties that dictate their environmental fate and bioavailability. Trace metals such as copper, nickel, zinc, and lead are inclined to interact with other elements and organic molecules present in the environment. Assessments of environmental fate and risks of trace metal exposure to wildlife and humans historically only accounted for these interactions superficially and are most commonly based on total metal concentrations. Thus, there is a need to develop methods and strategies to overcome this limitation and help improve risk assessments for both anthropogenic metal contamination, e.g., with copper, and geogenic background contamination, e.g., with nickel.

Copper reacts strongly with various functional groups present in soils and sediments, such as iron oxides and manganese oxides [1]. Localized deposits can be caused by anthropogenic activities such as mining, managing municipal and industrial waste, using copper as a pesticide in agriculture, and water treatment [2]. As a result, copper may enter the freshwater system naturally, as a result of human activity, or due to the corrosion of pipes and fittings in water distribution systems [3,4]. Copper is utilised in a wide variety of products, including pipes, paints, refinery oils, and construction materials, because of its malleability, conductivity, alloying potential, and resilience to corrosion and wear [2,4,5]. Copper exists in four different oxidation states: Cu (elemental copper), Cu (I) (cuprous ion), Cu (II) (cupric ion), and Cu (III) [6]. In water and sediment systems, Cu (II) is a species that is frequently found [7]. Since copper (I) is unstable, it easily oxidises to copper (II), which then undergoes additional redox processes and becomes hydrated in the presence of water molecules. The hydrated Cu (II) ion binds to dissolved organic and inorganic molecules as well as particulate debris [8]. In soil, copper does not often bioaccumulate or even hydrolyze in any form [9]. Cu speciation is complicated and dependent on a number of variables, including water chemistry (dissolved oxygen, pH, and redox potential), water hardness, and sediment–water interaction [10]. The toxicity of copper in freshwater is mainly driven by various chemical parameters and water quality. According to the biotic ligand model (BLM) proposed by [10], metal toxicity is driven by the accumulation of metal at a discrete site of action (biotic ligand). Metal speciation results in formation of inorganic and organic complexes [11]. As discussed above, Cu (II) adsorbs to particulate matter and forms dissolved complexes with both organic and inorganic ligands. This ability of Cu (II) is a major contributor for it to be the major contributor to copper toxicity (ECCC 2021). Water quality monitoring datasets from Environment and Climate Change Canada (ECCC) were compiled for different provinces and territories. Total copper concentrations of Canadian jurisdictions varied with an overall range from 0.002 (Ontario) to 5723 μg/L (Saskatchewan). The median copper concentrations ranged from 0.3 μg/L for Prince Edward Island (PEI) to 1.8 μg/L for Manitoba.

Nickel (Ni) is one of the most common elements on Earth and is widely distributed in the environment [12]. Ni occurs mostly in the mineral form at an average concentration of ~75,000 μg/kg in the Earth's crust. Ni is known to have a widespread distribution in the environment and is an essential component for many industrial and commercial uses. Various uses of nickel include electroplating, as a catalyst for fat hardening, coloring ceramics, electrical components, ammonia adsorption processes, and other metallurgical operations [13]. Nickel is released into the environment in various forms, predominantly from natural sources and few anthropogenic activities. Natural weathering and erosion of geological materials release nickel into surface waters and soils in Canada. Forest fires can be short-term but intense sources [14]. Mining, smelting, petroleum refining, and manufacturing industries are other major emitters [15]. Nickel moves in both particulate and dissolved forms in natural waters. The transport and fate of nickel in freshwater rely on various factors such as the pH, redox potential, ionic strength, type, and adsorption type [16]. Nickel toxicity is dependent upon the route of exposure and the solubility of a nickel compound [17]. The movement of nickel in the environment is high in acidic organic-rich soils, which can lead to groundwater contamination [18]. In sediments and suspended solids, most of the nickel is distributed among organic materials, precipitated and coprecipitated particle coatings, and crystalline particles.

Studying metal speciation in freshwater systems is challenging, as the applied methodologies need to be very sensitive to be able to recognize minute differences in relative proportions of metal species at trace concentration levels. Various models have been used to compare predictions of metal speciation to experimental values, which has been documented in a few studies [19–22]. Models such as the biotic ligand model (BLM), the Windermere humic aqueous model (WHAM), the NICA–Donnan model, and WASP—TOXI [23] have been developed to study or predict metal speciation in the environments at varying levels of complexity. BLMs determine metal speciation and predict metal toxicity to biota in aqueous systems using computational modelling. WHAM is based on the Humic Ion-Binding Model and assumes that proton and metal complexation occurs at two groups of discrete sites with strong and weak binding affinities [24]. The NICA–Donnan model considers carboxylic- and phenolic-type groups for determining site strengths [25]. In a study [26], modelled in situ concentrations of Cd, Cu, Ni, and Pb were compared with measured values by employing different speciation techniques, specifically the Humic Ion-Binding Model VI (WHAM 6) and the NICA–Donnan model. Both the models were found to be reasonably accurate and performed consistently for metal speciation in freshwater. However, concentrations of total dissolved Cu and Pb were underestimated by a large magnitude. While this finding is generally promising, the complexity of identifying anthropogenic versus geogenic sources of contamination, impacts of water chemistry/quality on the fate of metals, hydrology, and morphodynamics requires an integrated modelling strategy that accounts for all of these processes.

In this study, we used the Water Quality Analysis Simulation Program (WASP) to model the fate of the metals nickel (predominantly from geogenic background) and copper (predominantly from anthropogenic sources) between water and sediments in the South Saskatchewan River through the sampling periods of 2020 and 2021. With an average discharge of 277 m^3/s (min: 68 m^3/s, max: 731 m^3/s) in 2020 and 134 m^3/s (min: 68 m^3/s, max: 330 m^3/s) in 2021, the hydrology in these two sampling years was noticeably different (both measured at the outlet of Lake Diefenbaker). Additionally, Southern Saskatchewan experienced a unique pattern of precipitation in the two sampling years, with an average of 297.4 mm in 2020 and 180.7 mm in 2021 (ECCC historical weather data). This study makes use of trace metal concentrations and other physicochemical variables recently published by our group [27]. Comparing the predictive power of WASP for trace metal concentrations in sediment and water across two hydrologically different sampling periods, encompassing both a flood and a drought year, bears a considerable potential to test the model across a wide range of hydrologic conditions and, thus, the potential utility of this model in forecasting, e.g., with the goal of testing the impacts of climate change scenarios.

2. Materials and Methods

2.1. Study Site

The Saskatchewan River system, which originates in Alberta's Rocky Mountain headwaters and flows through Canada's plains, is made up of the South and North Saskatchewan Rivers. The South Saskatchewan River Basin, which includes the major cities of Saskatoon, Swift Current, Red Deer, Calgary, Lethbridge, and Medicine Hat, is situated in the southern regions of Alberta and Saskatchewan (Figure 1). Big Stick Lake, Bow River, Oldman River, Red Deer River, Seven Persons Creek, South Saskatchewan River, and Swift Current Creek are all parts of the South Saskatchewan River Basin [28].

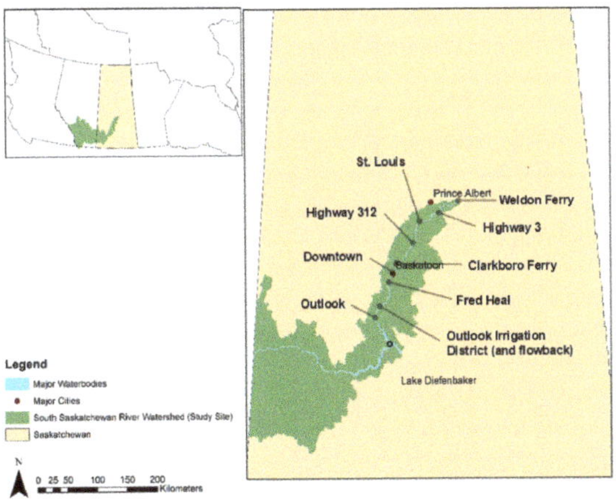

Figure 1. Sampling locations in the South Saskatchewan River Basin, Saskatchewan [27].

The Red Deer, Bow, and Oldman rivers meet at the confluence of the South Saskatchewan River, which is fed by the glaciers of the Rocky Mountains. From its source, the South Saskatchewan River flows for another 1392 kilometres. The SSR has an average discharge of 280 m^3/s at the Saskatchewan River Forks. It covers a watershed of approximately 146,100 km^2, of which 1800 km^2 are in Montana, United States, and 144,300 km^2 are in Alberta and Saskatchewan [29]. Agriculture accounts for two-thirds of the land cover [30].

2.2. Modelling Approach

The change of heavy metals in the South Saskatchewan River system was studied using the Water Quality Analysis Simulation Program 7.52 (WASP) coupled with HEC-RAS. The WASP model could be used to simulate many different aquatic systems. The application could simulate hazardous water contamination using the concepts of mass, momentum, and the conservation of energy. The hydrodynamics of the model domain was simulated using the hydraulic model HEC-RAS 6.3. Since HEC-RAS was developed by the U.S. Army Corp of Engineers (USACE), all users and organizations have access to a free version of the model. HEC-RAS is available to everyone and does not require a license, making it a key benefit of utilizing it as a modelling tool. HEC-RAS can model flow, sediment transport, water quality in one-dimensional (1D) steady and unsteady flow, and two-dimensional (2D) unsteady flow of rivers. The model uses geometric data representation and geometric and hydraulic computer algorithms for a network of river channels. The HEC-RAS program can simulate an input flood using either a (1D) unsteady flow model or a (2D) unsteady flow model following these parameters.

Flow data were acquired from available flow data from the Water Survey of Canada (WSC) station at Lake Diefenbaker, which has an operational gauge (records available from 1966 to the present). That station is about 13 km upstream of the modelling upstream boundary (gauge 07DA001—South Saskatchewan River Below Lake Diefenbaker). The HEC-RAS model was calibrated by modifying Manning's N to obtain the best match of the water level at 05HG001—South Saskatchewan River at Saskatoon and 05KD007—Saskatchewan River below the forks.

A 1D modelling approach was chosen as the optimum method for simulating the South Saskatchewan River sites. A 1D modelling approach was found to be reasonable for the region with optimum processing times. One-dimensional modelling can give results that are on par with, or better than, 2D models for rivers and floodplains where the predominant flow directions and forces follow the main river flow route, with less effort and fewer computing resources [31]. It was expected that the system was mixed both laterally and vertically. Previously, 1D WASP and HEC-RAS coupling [32] and quasi-2D [33,34] have been carried out.

HEC-RAS carried out the hydrodynamic part of the modelling, and the output of the HEC-RAS model was used as an input (for water depth at each segment) for the WASP model.

2.3. Setup for Water Quality Modelling

The water quality part of the study was modelled using the Water Quality Analysis Simulation Program (WASP 7.32). Due to some stability problems, the newer version of WASP was not considered for the current study. WASP was first created in the 1980s and has undergone numerous improvements [35]. The general dynamic model WASP uses a segmentation network to solve the conservation of momentum, energy, and mass equations and simulate the transport of contaminants and sediment. The WASP model is frequently utilized to address sediment transport [34], heavy metal transfer [33], and water quality issues [36,37]. The WASP stream transport module, TOXI, is coupled with flow routing for free-flow streams, ponded segments, and backwater reaches and is capable of calculating the flow of water, sediment, and dissolved constituents across branched and ponded segments. Additionally, model boundary constraints and input parameters are defined.

Water flows through a network of branching streams, which may contain both free-flowing and ponded parts, and is calculated using the standard WASP8 stream transport model. Flow routing can be estimated for free-flowing stream reaches, ponded reaches, and backwater or tidally-influenced reaches for one-dimensional branching streams or rivers. Advective transport can be driven through free-flowing portions with the simple yet practical kinematic wave flow routing method. The kinematic wave equation determines the propagation of flow waves and the fluctuations in flows, volumes, depths, and velocities that arise from the varied upstream inflow. This well-known equation is the solution to the one-dimensional continuity equation and a simplified form of the momentum equation that considers gravity and friction.

2.4. Kinematic Wave

For most simple river systems, the kinematic wave formulation is applicable. The kinematic wave calculates flow wave dissemination and flow velocity variations over a network of streams. The roughness of the slope and river bottom affect kinematic flow waves. By changing the Manning equation (Equation (1)) in the continuity equation and differentiating cross-sectional area with respect to time, the kinematic wave differential equation may be produced (Equation (2)).

$$Q = \frac{1}{n}\frac{A^{5/3}}{B^{2/3}}{S_0}^{1/2} \tag{1}$$

$$\frac{\partial Q}{\partial x} + \alpha\beta Q^{\beta-1}\frac{\partial Q}{\partial t} = 0 \tag{2}$$

where Q is the flow rate, x is longitudinal distance in channel, t is time, α and β are functions of hydraulic coefficients, S_0 is the slope of the channel, n is the Manning friction factor, A is the cross-sectional area, and B is the channel width. The depth exponent is used along with segment geometries to estimate hydraulic coefficients, which are later used to calculate segment flow depths under specific flow rates. [38] gave a complete description of WASP's stream transport. To quantify variations in velocities, widths, and depths across the network, the kinematic wave based on solutions of one-dimensional continuity equations, and a condensed version of the momentum equation that takes the effects of friction and gravity into account, was utilized.

2.5. Discretization

The river network was discretized using a 500 m interval. This led to 319 horizontal, one-dimensional water segments in WASP. Segment widths, average depths, and depth exponents were among the hydrodynamic metrics produced by HEC-RAS. The depth exponent controls channel shape in WASP. For these segments, a depth exponent value of 0.3 was used to simulate an uneven cross-section. Along with segment widths, slopes, and roughness factors, these depths are used in simulations to determine segment depths. The model's input variables include channel geometry, flow routing, boundary conditions, environmental time functions, loads, and initial segment conditions. Hydraulic parameters were obtained using a calibrated and validated HEC-RAS model and utilized as inputs for the flow functions in WASP [38]. From the 319 HEC-RAS cross-sections separated by 500 m, geometry for the WASP segments (average depths, widths, and slopes) was derived. Based on the station's water quality data closest to the study region, spatial linear interpolation was used to identify the boundary and initial conditions for the current investigation.

The study's duration was from 2020 to 2021. Initial heavy metal concentrations for each model component were derived by interpolating between two nearby long-term monitoring stations: the upstream boundary conditions were defined using the nearest long-term heavy metal data, which was Outlook. The boundary data for downstream was used based on the data from Weldon Ferry.

A segment length of 500 m was used for modelling. This length is deemed appropriate based on processing times, a reasonable uniform volume for segments, and an acceptable mixing. Benthic segments were inserted underneath each section of surface water to quantify erosion and sedimentation. The measured concentration at the upstream and downstream boundaries was used to calibrate the deposition rate of metals. In the HEC-RAS model, the spaces between each pair of cross-sections were split into individual WASP segments. Output files of the hydrodynamic model (HEC-RAS) were specific for each cross-section. These results then needed to be averaged between two consecutive cross-sections to obtain the exact values for each WASP segment. This procedure is shown in Figure 2 and was implemented in Microsoft Excel (Version 2211) for all input values.

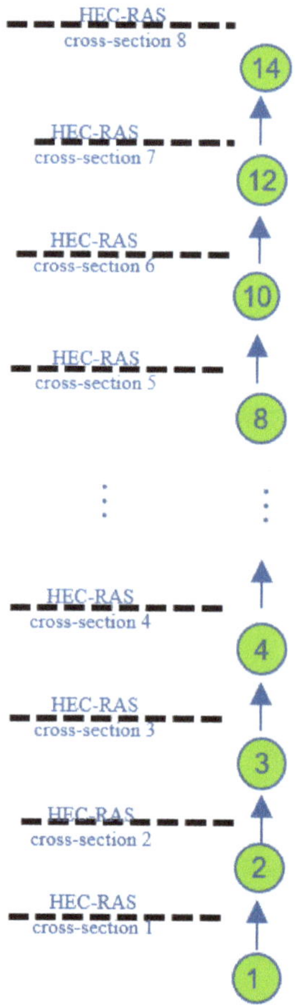

Figure 2. Model segmentation.

2.6. Visualization, Calibration, and Validation

The modelled heavy metal concentrations and hydrodynamics were calibrated and validated. The calibration of heavy metal concentrations involved several iterations in the parameter space within reasonable parameters of the observed values. The best curve fitting with the observed data was achieved by choosing the optimum parameter values. To validate the model, a run with these values was compared against an independent set of field data.

An effective way to review model simulations and calibrate them using collected data is to utilize the post-processor (MOVEM). Results from every WASP run, as well as others, can be visualized using MOVEM. MOVEM allows the modeller the choice of two graphical representations of the results: geometric grid and x/y plots. There is no restriction on how many geographical grids, x/y plots, or even model result files a user can employ in a session.

3. Results

3.1. Deposition Rates and Overall Concentration Trends

First, it was assumed that the deposition rate for copper and nickel should be the same as sediment. Deposition rates for copper and nickel were calibrated using the sampled concentration at the upstream and downstream model boundaries (Figure 3). A previously developed model for sediment transport was used, and the model was calibrated with an acceptable range of deposition [38]. Concentration plots (Figure 4) show the trend for both trace metals at selected sampling sites (Downtown, Clarkboro Ferry, and Highway 312) for both water and sediment samples. Concentration values for all sites can be found in the Supplementary Information.

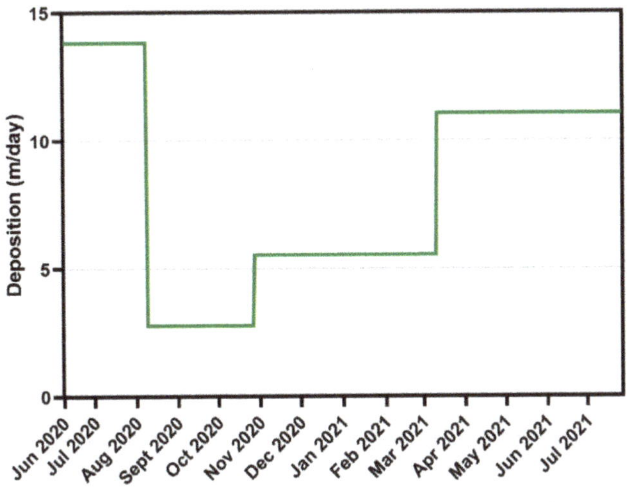

Figure 3. Calibrated deposition curve for both trace metals Cu and Ni.

Both trace metals were modelled for the same time periods for water and sediment in the years 2020 and 2021 and matched to the field sampling of water and sediment samples to maintain uniformity.

3.2. Copper

In 2020, copper concentrations were observed in the range of 1–5 µg/L except for two sites (Highway 3 and Weldon Ferry), with values at 42.2 µg/L and 20.2 µg/L, respectively (refer to Supplementary Information). Cu concentrations in river water in 2021 were in the range of 0.8 µg/L to 5 µg/L. That the observed concentrations remained uniform, for the most part, is indicative of the complex mixing of copper across the sampling sites, whereas elevated concentrations of copper at Highway 3 and Weldon Ferry could be attributed to various constructions around the location (bridges and roadway) in this area (S. Prajapati, pers. obs.). Many of the measured concentrations were over the long-term CCME water quality guideline for Cu of 4 µg/L at the given water hardness > 180 mg/L $CaCO_3$, which indicates that risks to aquatic life could not be excluded. Sediment Cu concentrations ranged from <LOQ to 19 mg/kg in 2020 and <LOQ to 15 mg/kg in 2021 [27]. Since the

resulting concentrations are below the CCME ISQG of 35.7 mg/kg, the authors do not expect adverse effects. In sediments, Cu ranged from 1–15 mg/kg, with Clarkboro Ferry being on the higher end for both sampling years, with values in the range of 12–15 mg/kg.

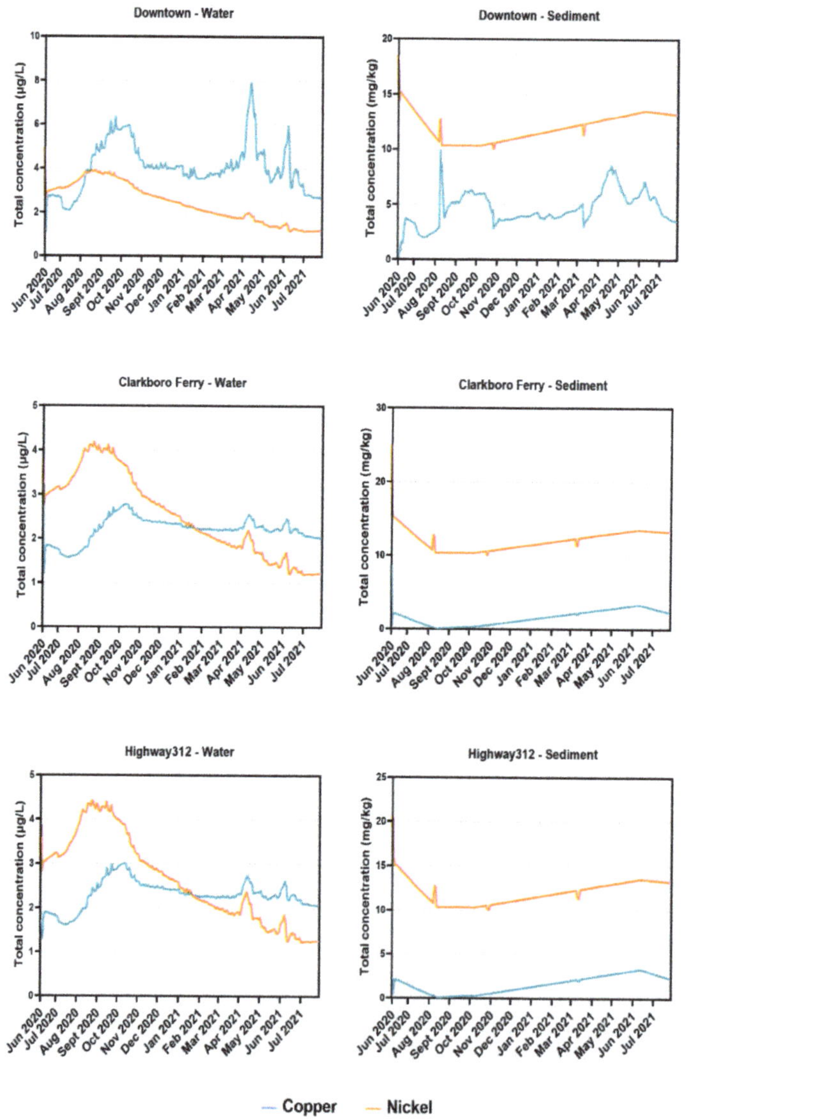

Figure 4. Concentrations of Cu and Ni in water and sediment for selected sites during the 2020 and 2021 field seasons.

Figures 5 and 6 show a longitudinal comparison of measured and modelled copper concentrations in water and sediment for 2020 and 2021. The concentrations modelled through the sampling time points almost uniformly increased between the upstream and downstream boundaries of the study area. Copper concentrations for water samples remained mostly consistent for June 2020, August 2020, and October 2020, but increased gradually and became more consistent in June 2021 and August 2021. Within the study period, discharge and flow velocity of the South Saskatchewan River were highest in June 2020, which can explain this observed pattern of copper concentrations, as increased discharge would cause significant erosion of riverbanks and stream beds. Thereafter, longitudinal profiles for August 2020, October 2020, and June 2021 remained relatively consistent throughout the segments.

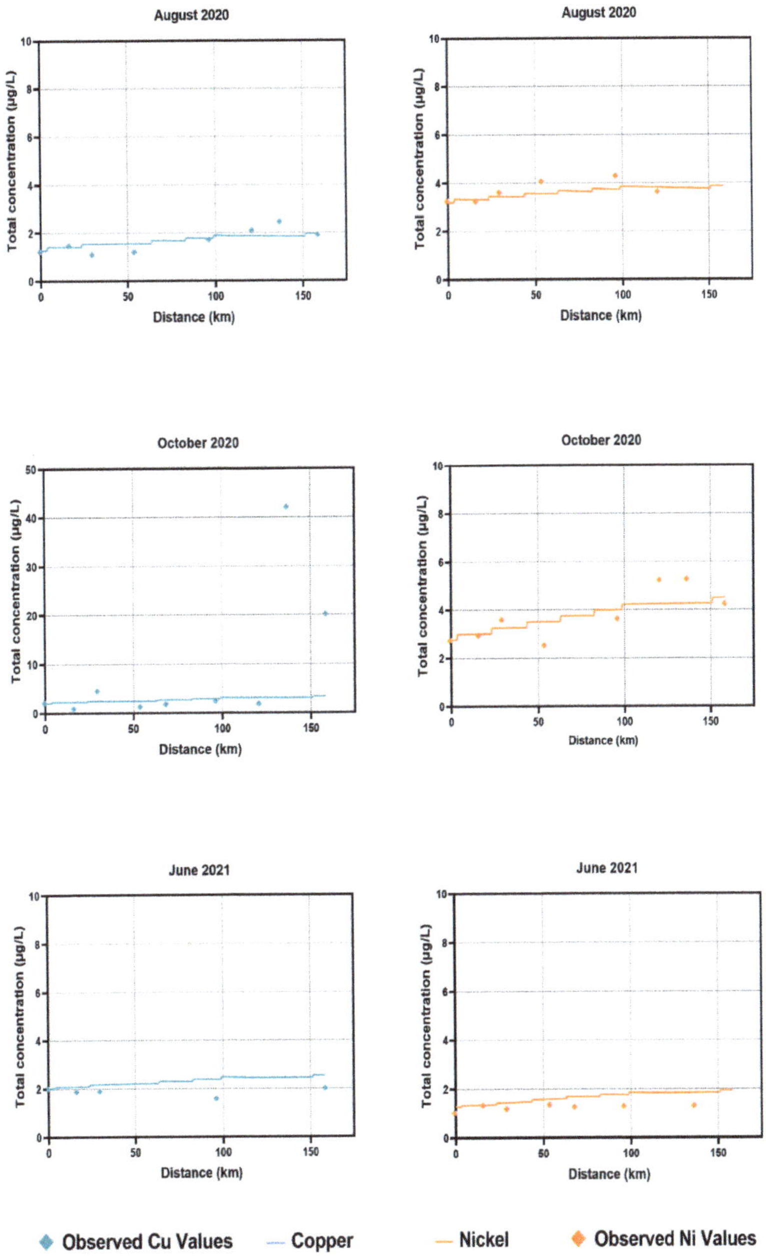

Figure 5. Longitudinal profiles of water concentrations with observed values for both trace metals in the 2020 and 2021 field seasons.

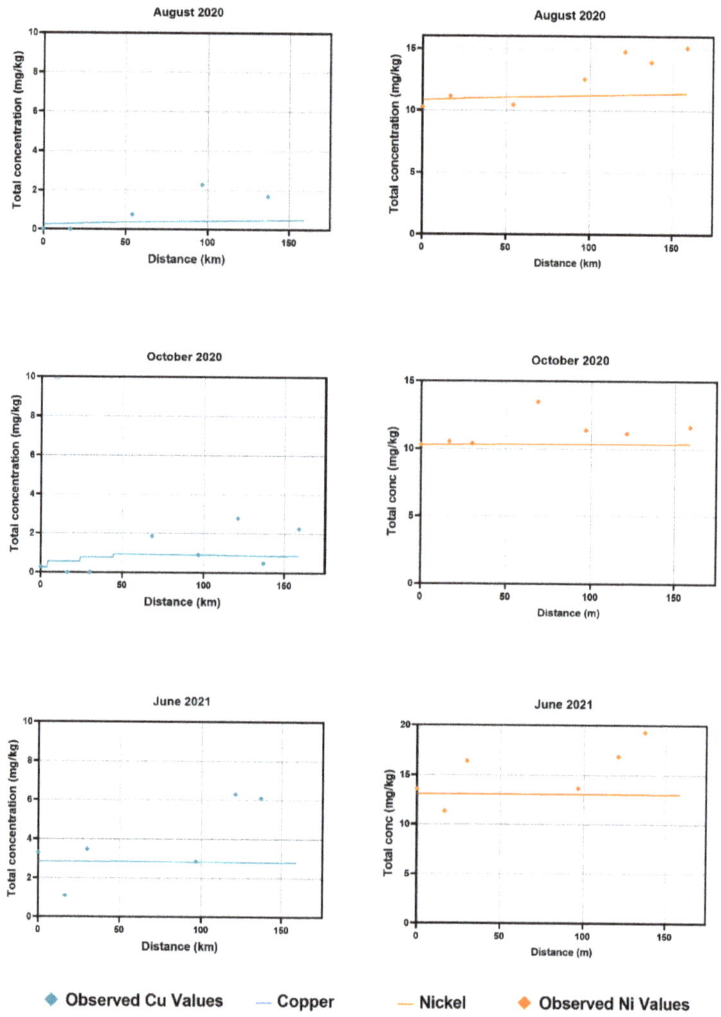

Figure 6. Longitudinal profiles of sediment concentrations for both trace metals in the 2020 and 2021 field seasons.

3.3. Nickel

In 2020, nickel concentrations ranged consistently from 2–6 μg/L for water samples and ranged from 0.3–1.4 μg/L in 2021. The high-flow season in 2020 (due to elevated discharge and precipitation mentioned above) helps explain the almost three- to fourfold greater nickel concentrations in that year. Levels of nickel in sediments for 2020 ranged from 10.5–25 mg/kg and mostly remained consistent throughout the year. However, in contrast to concentration values for water samples, the concentration of nickel in 2021 remained around the same range, with a very slight decrease, ranging from 10–20 mg/kg. No measured values were observed to be over the long-term CCME water quality guidelines for Ni of 150 μg/L at the given water hardness > 180 mg/L $CaCO_3$.

Longitudinal profiles for nickel (Figures 5 and 6) followed almost the same trend as those for copper, with very little variability in concentration values during the different sampling time points. Nickel concentrations remained consistent for August 2020, October 2020, June 2021, and August 2021, with a slight gradual increment with every major segment (i.e., sampling station). The concentration values followed a similar trend of gradual increase. For sediment samples, the modelled progressions again followed similar patterns in August 2020, October 2020, and June 2021, with a consistent and slight decline throughout the study area. Again, high flow in 2020 could explain the three- to fourfold greater nickel concentrations in 2020 compared to 2021 water samples. Similarly, for sediment samples, settling and transverse mixing can contribute to consistent and slight declines in the concentration values.

4. Discussion

Observed copper concentrations in water were consistent and followed similar trends to the observed values for most of the sampling time points. An exception to that is the observed values in the October 2020 sampling time points, where the copper concentration values were significantly higher than that of the trend following the WASP simulated values. This could be attributed to roadwork and bridge construction in the area, which would have been impossible for the model to cover. In sediment samples, the majority of the WASP simulated values were underestimated compared to the observed values, with a few exceptions. This could be the result of the continuous accumulation of metals in sediments over many years, which was not accounted for in the current modelling exercise.

Modelled nickel concentrations in water samples across 2020 and 2021 also followed a similar trend, without major discrepancies from the observed values. For sediment samples, there were a few instances where the observed values were higher than those of the modelled progression. The model tended to underestimate the nickel concentrations for sediment samples, following a similar trend to the copper concentrations for sediment samples.

During simulations, the model was found to be underestimating the concentrations at downstream segments. Additional diffuse loads were added to compensate for matching the simulated values with measured values. This significantly improved the accuracy for both metals. For water samples, additional loads (~1–3 kg/day) were included at major segments (sampling stations) as diffuse loads. The loads were increased 100-fold for the calibration of sediment samples. Calibration process determined these amounts. Copper and nickel are both known to sequester in the sediment bed and are less likely to be suspended in water. Copper, as mentioned above, enters the environment mainly through anthropogenic sources [4]. The release and remobilization of sediment-bound copper have been well-documented in various studies; however, the overall impacts are minor [39,40]. Mobility of nickel, on the other hand, is site-specific and depends on the soil type and pH [41]. These observations can be attributed to the higher observed values of nickel and copper in the sediment samples in the current study.

The model used in the current study is an extension of the previously developed model 1D WASP and HEC-RAS coupling [32] and quasi-2D [33,34]. In the study [38], the mixing of vanadium and sediment was modelled with a quasi-2D modelling approach in the lower Athabasca River. The authors observed a marked increase in vanadium concentration along the river. The model was successful in capturing the transverse mixing from tributary water with mainstream water.

As mentioned earlier, modelling freshwater systems is challenging, involving many complex processes that affect contaminants' transport and fate [42], which demands sensitive methods to obtain accurate results for different metals. With limited data availability, understanding metal speciation becomes difficult in such systems [43]. The current study for the South Saskatchewan River system serves as a preliminary study focusing on trace metal transport and fate. This study also provides an understanding of how the concentrations of trace metals (Cu and Ni) vary across upstream and downstream of the South Saskatchewan River. There were a few CCME ISQG exceedances for Cu concentrations, and there are no guidelines outlined for Ni (as of now). Although there may be no immediate danger just yet, contamination of aquatic systems cannot be excluded. Further study involving more metals in the model with sufficient input parameters would be an ideal recommendation.

5. Conclusions

The aim of this study was to better understand trace metal fate and interaction in the South Saskatchewan River. Two trace metals, copper (Cu) and nickel (Ni), were selected on the basis of their presence in the region. Copper is used in a wide range of products such as construction, pipes, industrial equipment, etc., resulting in widespread anthropogenic contamination with Cu. In contrast, nickel has a geogenic background in Saskatchewan and is transferred to the aquatic environment by effluents, leachates, and runoff from mining activities and the general land surface.

A 1D modelling approach was successfully applied in studying the trends and patterns of trace metal fate in the river and sediment. The simulations provided interesting data for both the trace metals and gave an insight into how the metals travel and react between water and sediment in the South Saskatchewan River.

A number of parameters needed to be assumed due to the lack of availability of consistent raw data for the modelling of these two trace metals. Improved availability of these parameter values would help make the model more efficient and accurate in predicting future trends with increased/decreased flows.

Supplementary Materials: The following supporting information can be downloaded at: https://www.mdpi.com/article/10.3390/w15020265/s1, Trace metal concentrations can be found in the spreadsheet attached Table S1: Heavymetal concentrations.xlsx.

Author Contributions: Conceptualization, S.P., P.S., M.B. and K.-E.L.; methodology, S.P. and P.S.; software, S.P. and P.S.; writing—original draft preparation, S.P. and P.S.; writing—review and editing, M.B. and K.-E.L.; supervision, M.B. and K.-E.L. All authors have read and agreed to the published version of the manuscript.

Funding: This research was funded University of Saskatchewan's New Faculty Graduate Student Support Program (NFGSSP), Global Water Futures (GWF), Canada First Research Excellence Funds, and the Canada Foundation for Innovation (CFI)—38581.

Data Availability Statement: All the research data supporting reported results can be found in the Supplementary Materials.

Conflicts of Interest: The authors declare no conflict of interest.

References

1. Tonkin, J.W.; Balistrieri, L.S.; Murray, J.W. Modeling sorption of divalent metal cations on hydrous manganese oxide using the diffuse double layer model. *Appl. Geochem.* **2004**, *19*, 29–53. Available online: http://pubs.er.usgs.gov/publication/70027390 (accessed on 18 October 2022). [CrossRef]
2. ATSDR. Toxicological Profile for Copper. Public Health 2004. Volume 314. Available online: https://www.atsdr.cdc.gov/ToxProfiles/tp132.pdf%0Ahttps://stacks.cdc.gov/view/cdc/7006 (accessed on 18 October 2022).
3. Food, T. *Copper in Drinking Water*; National Research Council: Rockville, MD, USA, 2000.
4. Ellingsen, D.; Horn, N.; Aaseth, J. Chapter 26—Essential metals: Assessing risks from deficiency and toxicity. In *Handbook on the Toxicology of Metals*, 3rd ed.; Academic Press: Cambridge, MA, USA, 2007; pp. 163–176.
5. WHO. *WHO Guidelines for Drinking-Water Quality*; Eisei kagaku; WHO: Geneva, Switzerland, 1998; Volume 35, pp. 307–312.
6. Georgopoulos, P.G.; Roy, A.; Yonone-Lioy, M.J.; Opiekun, R.E.; Lioy, P.J. Environmental copper: Its dynamics and human exposure issues. *J. Toxicol. Environ. Health B Crit. Rev.* **2001**, *4*, 341–394. [CrossRef] [PubMed]
7. Gaetke, L.M.; Chow-Johnson, H.S.; Chow, C.K. Copper: Toxicological relevance and mechanisms. *Arch. Toxicol.* **2014**, *88*, 1929–1938. [CrossRef] [PubMed]
8. Health Canada. *Copper Pesticides*; Health Canada: Ottawa, ON, Canada, 2009.
9. Cuppett, J.D.; Duncan, S.E.; Dietrich, A.M. Evaluation of copper speciation and water quality factors that affect aqueous copper tasting response. *Chem. Senses* **2006**, *31*, 689–697. [CrossRef]
10. Pagenkopf, G.K.; Russo, R.C.; Thurston, R.V. Effect of complexation on toxicity of copper to fishes. *J. Fish. Res. Board. Can.* **1974**, *31*, 462–465. [CrossRef]
11. Santore, R.C.; Di Toro, D.M.; Paquin, P.R.; Allen, H.E.; Meyer, J.S. Biotic ligand model of the acute toxicity of metals. 2. Application to acute copper toxicity in freshwater fish and Daphnia. *Environ. Toxicol. Chem.* **2001**, *20*, 2397–2402. [CrossRef] [PubMed]

12. Cempel, M.; Nikel, G. Nickel: A review of its sources and environmental toxicology. *Pol. J. Environ. Stud.* **2006**, *15*, 375–382.
13. IPCS. *Environmental Health Criteria*; Environ Heal Criteria: Geneva, Switzerland, 1991; Volume 204.
14. Havas, M.; Hutchinson, T.C. The Smoking Hills: Natural acidification of an aquatic ecosystem. *Nature* **1983**, *301*, 23–27. [CrossRef]
15. Environment Canada. National Pollutant Release Inventory. 2007. Available online: https://www.canada.ca/en/services/environment/pollution-waste-management/national-pollutant-release-inventory.html (accessed on 24 October 2022).
16. Callahan, M.A. *Water Related Environmental Fate of 129 Priority Pollutants*; USEPA: Washington, DC, USA, 1979; I:43-1-43–10.
17. Coogan, T.P.; Latta, D.M.; Snow, E.T.; Costa, M.; Lawrence, A. Toxicity and Carcinogenicity of Nickel Compounds. *CRC Crit. Rev. Toxicol.* **1989**, *19*, 341–384. [CrossRef]
18. Nieminen, T.M.; Ukonmaanaho, L.; Rausch, N.; Shotyk, W. Biogeochemistry of Nickel and Its Release into the Environment. *Met. Ions Life Sci.* **2007**, *2*, 1–30.
19. Bryan, S.E.; Tipping, E.; Hamilton-Taylor, J. Comparison of measured and modelled copper binding by natural organic matter in freshwaters. *Comp. Biochem. Physiol. C Toxicol. Pharmacol.* **2002**, *133*, 37–49. [CrossRef] [PubMed]
20. Dwane, G.C.; Tipping, E. Testing a humic speciation model by titration of copper-amended natural waters. *Environ. Int.* **1998**, *24*, 609–616. Available online: https://www.sciencedirect.com/science/article/pii/S0160412098000464 (accessed on 3 November 2022). [CrossRef]
21. Christensen, J.B.; Christensen, T.H. Complexation of Cd, Ni, and Zn by DOC in Polluted Groundwater: A Comparison of Approaches Using Resin Exchange, Aquifer Material Sorption, and Computer Speciation Models (WHAM and MINTEQA2). *Environ. Sci. Technol.* **1999**, *33*, 3857–3863. [CrossRef]
22. Oste, L.A.; Temminghoff, E.J.M.; Lexmond, T.M.; Van Riemsdijk, W.H. Measuring and modeling zinc and cadmium binding by humic acid. *Anal. Chem.* **2002**, *74*, 856–862. [CrossRef] [PubMed]
23. Ambrose, R.B.; Wool, T.A. WASP7 Stream Transport—Model Theory and User 's Guide Supplement to Water Quality Analysis Simulation Program (WASP) User WASP7 Stream Transport—Model Theory and User ' s Guide Supplement to Water Quality Analysis. *Environ. Prot.* **2009**.
24. Tipping, E.; Hurley, M.A. A unifying model of cation binding by humic substances. *Geochim. Cosmochim. Acta* **1992**, *56*, 3627–3641. Available online: https://www.sciencedirect.com/science/article/pii/001670379290158F (accessed on 3 November 2022). [CrossRef]
25. Kinniburgh, D.G.; Milne, C.J.; Benedetti, M.F.; Pinheiro, J.P.; Filius, J.; Koopal, L.K.; Van Riemsdijk, W.H. Metal Ion Binding by Humic Acid: Application of the NICA-Donnan Model. *Environ. Sci. Technol.* **1996**, *30*, 1687–1698. [CrossRef]
26. Sigg, L.; Black, F.; Buffle, J.; Cao, J.; Cleven, R.; Davison, W.; Galceran, J.; Gunkel, P.; Kalis, E.; Kistler, D.; et al. Comparison of Analytical Techniques for Dynamic Trace Metal Speciation in Natural Freshwaters. *Environ. Sci. Technol.* **2006**, *40*, 1934–1941. [CrossRef]
27. Prajapati, S.; Challis, J.K.; Jardine, T.D.; Brinkmann, M. Levels of pesticides and trace metals in water, sediment, and fish of a large, agriculturally-dominated river. *Chemosphere* **2022**, *308 Pt 1*, 136236. [CrossRef]
28. Lac, S.; Colan, C. *South Saskatchewan River Basin Biogeography*; Institutional Adaptation and Transformation in Climate Change: Regina, SK, Canada, 2004; pp. 1–44.
29. Alberta, E. South Saskatchewan River Basin. *Reg. Serv. South Reg.* **2003**, *2003*, 1–39.
30. Rosenberg, D.; Chambersy, P.; Culp, J.M.; Franzin, W.G.; Nelson, P.A.; Salki, A.G. *Rivers of North America*; Chapter 19; Benke, A., Cushing, C., Eds.; Association of American Publishers: Washington, DC, USA, 2005; pp. 853–901.
31. US Army Corps of Engineers. HEC-RAS River Analysis System: 1D Vs. 2D Hydraulic Modeling. HEC-RAS River Anal Syst User's Man 2022. Volume 6.3, pp. 1–4. Available online: https://www.hec.usace.army.mil/confluence/rasdocs/r2dum/latest/steady-vs-unsteady-flow-and-1d-vs-2d-modeling/1d-vs-2d-hydraulic-modeling (accessed on 6 November 2022).
32. Akomeah, E.; Lindenschmidt, K.-E. Seasonal Variation in Sediment Oxygen Demand in a Northern Chained River-Lake System. *Water* **2017**, *9*, 254. Available online: https://www.mdpi.com/2073-4441/9/4/254 (accessed on 14 October 2022). [CrossRef]
33. Lindenschmidt, K.E.; Sabokruhie, P.; Rosner, T. Modelling transverse mixing of sediment and vanadium in a river impacted by oil sands mining operations. *J. Hydrol. Reg. Stud.* **2022**, *40*, 101043. [CrossRef]
34. Sabokruhie, P.; Akomeah, E.; Rosner, T.; Lindenschmidt, K.E. Proof-of-concept of a quasi-2d water-quality modelling approach to simulate transverse mixing in rivers. *Water* **2021**, *13*, 3071. [CrossRef]
35. Wool, T.; Ambrose, R.B.; Martin, J.L.; Comer, A. WASP 8: The Next Generation in the 50-year Evolution of USEPA's Water Quality Model. *Water* **2020**, *12*, 1398. Available online: https://www.mdpi.com/2073-4441/12/5/1398 (accessed on 15 July 2022). [CrossRef]
36. Mamani Larico, A.J.; Zúñiga Medina, S.A. Application of WASP model for assessment of water quality for eutrophication control for a reservoir in the Peruvian Andes. *Lakes Reserv. Res. Manag.* **2019**, *24*, 37–47. [CrossRef]
37. Shabani, A.; Woznicki, S.A.; Mehaffey, M.; Butcher, J.; Wool, T.A.; Whung, P.Y. A coupled hydrodynamic (HEC-RAS 2D) and water quality model (WASP) for simulating flood-induced soil, sediment, and contaminant transport. *J. Flood Risk Manag.* **2021**, *14*, 1–17. [CrossRef] [PubMed]
38. Costello, D.M.; Hammerschmidt, C.R.; Burton, G.A. Copper Sediment Toxicity and Partitioning during Oxidation in a Flow-Through Flume. *Environ. Sci. Technol.* **2015**, *49*, 6926–6933. [CrossRef]
39. Teuchies, J.; Bervoets, L.; Cox, T.J.S.; Meire, P.; de Deckere, E. The effect of waste water treatment on river metal concentrations: Removal or enrichment? *J. Soils Sediments* **2011**, *11*, 364–372. [CrossRef]

40. De Jonge, M.; Teuchies, J.; Meire, P.; Blust, R.; Bervoets, L. The impact of increased oxygen conditions on metal-contaminated sediments part I: Effects on redox status, sediment geochemistry and metal bioavailability. *Water Res.* **2012**, *46*, 2205–2214. [CrossRef] [PubMed]
41. Richter, R.; Theis, T.R. *Nickel Speciation in a Soil/Water System*; John Wiley & Sons: Hoboken, NJ, USA, 1989.
42. Rahimpour Asenjan, M.; Danesh-Yazdi, M. The effect of seasonal variation in precipitation and evapotranspiration on the transient travel time distributions. *Adv. Water Resour.* **2020**, *142*, 103618. [CrossRef]
43. Unsworth, E.R.; Warnken, K.W.; Zhang, H.; Davison, W.; Black, F.; Buffle, J.; Cao, J.; Cleven, R.; Galceran, J.; Gunkel, P.; et al. Model predictions of metal speciation in freshwaters compared to measurements by in situ techniques. *Environ. Sci. Technol.* **2006**, *40*, 1942–1949. [CrossRef] [PubMed]

water

MDPI

Article

Discrimination of Chemical Oxygen Demand Pollution in Surface Water Based on Visible Near-Infrared Spectroscopy

Xueqin Han [1,†], Xiaoyan Chen [2,†], Jinfang Ma [1], Jiaze Chen [1], Baiheng Xie [1], Wenhua Yin [2], Yanyan Yang [2], Wenchao Jia [2], Danping Xie [2,*] and Furong Huang [1,*]

1 Opto-Electronic Department, Jinan University, Guangzhou 510632, China
2 State Environment Protection Key Laboratory of Water Environmental Simulation and Pollution Control, South China Institute of Environmental Sciences, Ministry of Ecology and Environment, Guangzhou 510655, China
* Correspondence: xiedanping@scies.org (D.X.); furong_huang@jnu.edu.cn (F.H.)
† These authors contributed equally to this work.

Abstract: Chemical oxygen demand (COD) is one of the indicators used to monitor the level of pollution in surface water. To recycle agricultural water resources, it is crucial to monitor, in a timely manner, whether COD in surface water exceeds the agricultural water control standard. A diagnostic model of surface water pollution was developed using visible near-infrared spectroscopy (Vis-NIR) combined with partial least squares discriminant analysis (PLS–DA). A total of 127 surface water samples were collected from Guangzhou, Guangdong, China. The COD content was measured using the potassium dichromate method. The spectra of the surface water samples were recorded using a Vis-NIR spectrometer, and the spectral data were pre-processed using four different methods. To improve the accuracy and simplicity of the model, the synthetic minority oversampling technique (SMOTE) and the competitive adaptive reweighted sampling (CARS) algorithm were used to enhance model performance. The best PLS–DA model achieved an accuracy of 88%, and the SMOTE–PLS–DA model had an accuracy of 94%. The SMOTE algorithm could improve the accuracy of the model despite the sampling imbalance. The CARS–SMOTE–PLS–DA model achieved 97% accuracy, and the CARS band selection technique improved the simplicity and accuracy of the discrimination model. The CARS–SMOTE–PLS–DA model improved the discrimination accuracy by 9% over that of the PLS–DA model. This method can not only save human and material resources but is also a new way for real-time online discrimination of COD in surface water.

Keywords: surface water; vis-NIR spectroscopy; chemical oxygen demand; SMOTE; CARS; PLS–DA

Citation: Han, X.; Chen, X.; Ma, J.; Chen, J.; Xie, B.; Yin, W.; Yang, Y.; Jia, W.; Xie, D.; Huang, F. Discrimination of Chemical Oxygen Demand Pollution in Surface Water Based on Visible Near-Infrared Spectroscopy. *Water* **2022**, *14*, 3003. https://doi.org/10.3390/w14193003

Academic Editor: Karl-Erich Lindenschmidt

Received: 24 August 2022
Accepted: 20 September 2022
Published: 23 September 2022

1. Introduction

Sustainable development of the ecological environment is the common demand for the survival of all mankind, and the recycling of agricultural water resources is urgent. Surface water is one of the main sources of water for agricultural irrigation and an important factor affecting the quality of crops [1,2]. With the rapid advancement of industrialization in modern society, the random discharge of industrial wastewater has become an increasingly serious environmental problem. Extensive domestic garbage and industrial chemical residues flow into surface water, resulting in the deposition of a variety of harmful chemicals. This poses a serious threat to the recycling of agricultural water resources [3,4]. Accurate judgment of surface water pollution is one of the means to ensure the quality of agricultural cultivation.

Surface water pollutants are mainly organic and are generally quantitatively indicated by the chemical oxygen demand (COD). Conventional methods to test COD include the dichromate method and the permanganate index method. These methods not only require chemical reagents but also have the shortcomings of complex chemical reactions and long time periods. Moreover, they are likely to cause secondary pollution if the chemical reagents are not handled properly [5]. Hence, to protect and recycle surface water resources, it is necessary to develop a rapid, effective, and eco-friendly detection technology to accurately evaluate the degree of surface water pollution [6].

Visible near-infrared spectroscopy (Vis-NIR) is a green, non-destructive, and rapid detection technique. It is widely used in fields such as the ecological environment and medicine through a combination of statistical modeling and chemometric methods [7]. This analytical technique not only provides rich qualitative and quantitative information on substances but also has the advantages of being non-destructive and easy to apply. Therefore, this technique is widely used to detect various water pollution indicators [8]. The analytical method of this technique mainly involves the establishment of a calibration model using the spectra and conventional values of the target components. Linear discriminant models such as partial least squares discriminant analysis (PLS–DA) are commonly used in spectral modeling owing to their simple structure and ease of operation [9]. PLS–DA is a classification technique based on partial least squares. Its mathematical basis is principal component analysis, and the regression model between the independent variable and the categorical variable of the training sample is mainly established by the information of the samples in the process of features selection, and then the characteristic variables related to the classification are effectively extracted [10].

On the one hand, the accuracy and stability of the model will be affected by less representative sample data and the skewed distribution of sample categories [11]. Uneven distribution can easily occur when collecting samples. Therefore, the key factor affecting the performance of the classification model is the quantity distribution of the samples in different categories. Common machine learning algorithms adopt a balanced training set, where all categories are represented equally [12]. However, such treatment leads to a certain error in the prediction of categories with a large number of samples, whereas categories with a small number of samples are prone to misclassification [13]. On the other hand, the accuracy and stability of the model will suffer from redundancy in the spectral data. If the entire Vis-NIR band is used to train the model, it is often too complex and may produce inefficient models [14]. Some spectral variables may contain irrelevant or even noise information, which may distort the true relationship between the sample information and Vis-NIR predictors. Spectral selection algorithms are applied to overcome the drawbacks of spectral analysis. The competitive adaptive reweighted sampling (CARS) algorithm is one of the most commonly used band selection strategies [15]. This algorithm eliminates unimportant spectral variables when extracting the optimal subset of such variables according to the regression coefficients. However, it has not been validated whether this algorithm can effectively discriminate if the COD of surface water exceeds the threshold through Vis-NIR.

In our last article, we achieved quantitative predictions for surface water, but not very good predictions for COD greater than 120 mg/L [16]. In this experiment, samples that were more seriously polluted and whose COD was greater than 600 mg/L were added, and the method of qualitative discrimination was tried to achieve high-accuracy COD online discrimination, which provided new ideas for surface water quality management.

The purpose of this study was to explore the best comprehensive modeling approach of Vis-NIR to diagnose whether the COD of surface water exceeds its management value. The following objectives were considered: (1) to understand the effect of spectral pre-processing methods on the discrimination results of surface water COD; (2) to improve the distribution of sample categories using the synthetic minority oversampling technique (SMOTE) algorithm; (3) to develop a CARS-SMOTE-PLS-DA model for rapid determination of COD in surface water using the CARS band selection algorithm and the SMOTE algorithm; and (4) to determine the important wavelengths for the discrimination of surface water COD and the relevant components of surface water pollution.

2. Materials and Methods

2.1. Study Area and Sample Collection

The samples for this study were provided by the South China Research and Monitoring Analysis Center, South China Institute of Environmental Sciences, Ministry of Ecology and Environment. These samples were from Guangzhou, Guangdong, China. Surface water was collected from an inland river in Guangzhou that was often used as the water source for agricultural irrigation and the daily life of residents. A total of 127 samples were collected from 15 July to 15 October 2021. They were placed in sealed test tubes and labeled in the sampling order, and then delivered to the laboratory at room temperature on 16 October 2021. The COD value of each sample was determined in all experiments using the conventional permanganate index method [17]. The measured COD values were used for the calibration and validation of spectral analysis.

2.2. Chemical Analysis and Contamination Assessment

To determine the COD content of surface water, a known amount of potassium dichromate solution was added to 127 surface water samples with silver salts as the catalyst in a strong acid medium. After boiling and refluxing, the unreduced potassium dichromate in the samples was titrated with ferrous ammonium sulfate using the ferroin indicator solution as the indicator. The mass concentration of oxygen consumed was calculated based on the amount of potassium dichromate consumed, which was the specific value of COD.

The collected surface water samples were divided into two categories according to the COD threshold value (40 mg/L) required for Class V in the environmental quality standards for surface water (GB3838-2002), which is applicable to surface water for agricultural use and in the general landscape. They were further coded as binary 0 or 1 to indicate the COD content of each water sample as below or above the threshold, respectively [18].

2.3. Spectrum Acquisition and Pre-Processing

An XDS Rapid Content liquid grating Vis-NIR spectrometer (with a transmission attachment and a 2 mm quartz cuvette) from FOSS, Denmark was used for this study. The spectra were collected in the range of 400–2500 nm, including most of the NIR region. The wavelength sampling interval was 2 nm. An appropriate amount of sample was placed in a quartz cuvette with an optical path of 2 mm to collect the spectra. Each sample was scanned thrice in the NIR spectrometer. Then, the average spectrum of the three scans was taken as the acquisition spectrum. At the end of each acquisition, the quartz cuvette was cleaned with distilled water and dried with filter paper. The spectrum of each sample was measured at room temperature (24 ± 1 °C) and humidity ($46\% \pm 1\%$ RH).

The measured spectrum was inevitably affected by instrument noise and the ambient environment. Therefore, four spectral pre-processing methods were used for the spectra of the water samples: first derivative (FD), second derivative (SD), multiplicative scatter correction (MSC), and standard normal variate (SNV) [19].

2.4. Sample Set Partitioning and Model Evaluation Parameters

The sample set partitioning based on joint X-Y distance (SPXY) [20,21] was used. The training and test sets were partitioned with a ratio of 3:1. The training set could identify different classes of spectral patterns upon fitting the classification model, whereas the test set was used to evaluate the performance of the model. The specific partitioning results with the surface water sample information are shown in Table 1.

Table 1. Statistics of surface water COD values and partitioning of sample set.

Sample Type	Set	The Range of PH	Number of Samples	Min (mg/L)	Max (mg/L)	Mean (mg/L)	Median (mg/L)	COD Value > 40 mg/L	COD Value < 40 mg/L
Surface water	All	5.63–8.92	127	4	688	61.98	27	39	88
	Training set	5.63–7.85	95	4	688	58.65	20	25	70
	Testing set	6.52–8.92	32	5	313	50.25	18	14	18

Notes: COD: chemical oxygen demand; Min: minimum; Max: maximum.

2.5. Evaluation of the Model Performance

The accuracy, sensitivity, and specificity were used to evaluate the overall performance of the classification models. The classification accuracy refers to the ratio of the number of samples correctly discriminated to the total number of samples in the classification model when testing the established model using the prediction set. Sensitivity and specificity are two key metrics for the classification model that indicate the percentage of positive and negative samples correctly classified, respectively. When the accuracy, specificity, and sensitivity are closer to 1, the classification model has better performance.

$$\text{Accuracy} = \frac{TP+TN}{TP+FP+TN+FN} \quad (1)$$

$$\text{Sensitivity} = \frac{TP}{TP+FN} \quad (2)$$

$$\text{Specificity} = \frac{FP}{FP+TN} \quad (3)$$

where TP denotes the number of positive samples in the predication set correctly classified by the mode; FN denotes the number of positive samples in the test set incorrectly classified by the model; FP denotes the number of negative samples in the test set incorrectly classified by the model; and TN denotes the number of negative samples in the test set correctly classified by the model.

2.6. Synthetic Minority Oversampling Technique

When modeling algorithms are applied directly to data with uneven and unbalanced sample distribution, samples in the categories with smaller quantities are prone to misclassification, which reduces the overall accuracy [22]. Therefore, improving the discrimination accuracy of minority samples in discriminant analysis is a key issue. The numbers of collected surface water samples with COD values below and above the threshold were quite different; the number of samples with COD values above the threshold was small. Therefore, the oversampling method of the SMOTE algorithm was used to improve the sample distribution. New surface water categories were generated in the training set such that below-threshold and above-threshold samples obtained balanced observations (equal number of samples per category in the training set). The SMOTE algorithm proposed by Chawla et al. [23] is an efficient oversampling technique that can be used to avoid the overfitting that arises from the direct replication of a small quantity of samples. The SMOTE technique runs the oversampling difference by introducing synthetic examples into the spectral space and adding K-nearest neighbors. The K value was set to five to control the newly generated examples. For the original dataset, the sample training set corresponding to each pre-processing method generated a SMOTE-processed training set.

2.7. Competitive Adaptive Reweighted Sampling

CARS is a wavelength selection method that adopts the Darwinian evolution theory of "survival of the fittest". The key wavelengths selected are those with relatively large absolute coefficients in the multiple linear regression model. This selection method conducts a wavelength selection based on the exponential decay function (EDF) and then selects the key wavelengths based on the competitive wavelength selection of adaptive reweighted sampling [24,25]. The algorithm implementation is divided into the following four steps:

(1) Perform monte carlo sampling and select a certain proportion of samples to build a calibration model;
(2) Use EDF to remove the number of wavelengths with low absolute values of regression coefficients;
(3) Calculate the root mean square error cross-validation (RMSECV) and filter out significant wavelengths using adaptive reweighted sampling (ARS);
(4) Select the subset with the lowest RMSECV as the best wavelength combination.

EDF can realize the rapid elimination and selection of wavelengths. In each sampling process, the wavelength ratio to be retained is calculated by using EDF. The calculation formula of the wavelength ratio is as follows:

$$r_i = ae^{ki} \tag{4}$$

Among them, a is related to two fast constants, which are related to the number of spectral wavelengths p and the number of sampling runs N in CARS.

$$a = \left(\frac{p}{2}\right)^{\frac{1}{N-1}} \tag{5}$$

$$k = \frac{\ln\left(\frac{p}{2}\right)}{N-1} \tag{6}$$

After forced wavelength reduction by EDF, ARS is used to imitate the principle of survival of the fittest, and wavelengths are eliminated in a competitive manner. In ARS, variables will be randomly weighted and sampled, and variables with larger weights will be selected.

The aforementioned algorithms were run in MATLAB (R2018a, Math Works, Inc., Natick, MA, USA).

3. Results and Discussion

3.1. Descriptive Statistics

The statistics of the COD values for surface water samples measured in the laboratory and the partitioning of the sample set are shown in Table 1. The PH of these samples ranges from 5.63 to 8.92. The COD values of all surface water samples varied between 4 and 688 mg/L, with a mean value of 61.98 mg/L and a median value of 27 mg/L. It was also evident that the number of samples with a COD value below the threshold value far exceeded that of samples with a COD value above the threshold value. Samples with a COD value larger than the threshold value may be influenced by human activities and natural factors related to landscape changes, such as domestic wastewater discharge. The training and test sets were divided using the SPXY method with a ratio of 3:1. A total of 95 and 32 samples in the dataset were divided into the training and test sets, respectively. The results of the division are shown in Table 1.

Figure 1a shows the distribution of COD content of surface water samples. Most COD values were between 1 and 100 mg/L. Samples with COD values greater than 40 mg/L were designated as contaminated samples and those with COD values less than 40 mg/L were uncontaminated samples. A total of 25 samples in the training set were contaminated and 70 samples were uncontaminated, whereas 14 samples in the test set were contaminated and 18 samples were uncontaminated. This indicated a large gap between the number of uncontaminated and contaminated samples in the training set, which was likely to affect the modeling results. Therefore, in the subsequent analysis, the SMOTE algorithm was used to generate new surface water categories in the training set, so that the uncontaminated and contaminated samples achieved balanced observations. The impact of the excessive gap between the number of categories on the modeling was examined by comparing this with the modeling results without using the SMOTE algorithm. Additionally, the feasibility of the SMOTE algorithm was verified. With the application of the SMOTE algorithm, the uncontaminated samples (<40 mg/L) in the training set increased from 25 to 70, forming a new training set. The numbers of uncontaminated and contaminated samples in the training set are shown in Figure 1b.

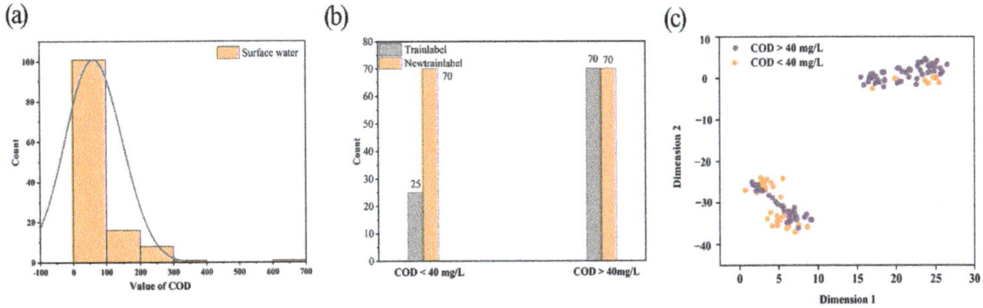

Figure 1. (**a**) Histogram of surface water COD value distribution, (**b**) distribution of training set samples before and after SMOTE expansion, (**c**) t−SNE visualization results plot.

The Vis-NIR spectral data were mapped into two-dimensional space using the t-distributed stochastic neighbor embedding (t-SNE) visualization algorithm. Thus, the variability and inherent characteristics of the Vis-NIR spectral datasets of the uncontaminated and contaminated samples could be understood more intuitively. During visualization, the t-SNE method can preserve the nonlinear structure of the spectral dataset [26]. In contrast to principal component analysis, t-SNE searches for the data structure based on a random probability distribution over the domain graph [27]. The visualization results of the surface water sample dataset upon using the t-SNE algorithm are shown in Figure 1c. The dataset forms two distinct clusters, wherein each point represents a sample, and the axes represent the first two dimensions of the t-SNE features. These t-SNE visualization results further validate the feasibility of using the Vis-NIR spectral technique for discriminant analysis of surface water COD.

3.2. Spectral Absorption Characteristics

Figure 2a–d present the original spectra of surface water and those after SD pre-processing for 400–2500, 1200–1500, and 1800–2200 nm, respectively. In Figure 2a, the spectra of uncontaminated and contaminated samples show similar trends and shapes. However, after SD pre-processing, the spectra show multiple peaks and troughs. Since there are large peaks and troughs near 1800 nm, the spectra after SD pre-processing were locally amplified to obtain Figure 2c,d. These figures show more pronounced absorptions at 1400, 1450, and 1980 nm, which may be caused by the stretching vibrations of the O-H, C-H, and N-H bonds, respectively [28–30]. They also show that the uncontaminated and contaminated samples exhibited large differences in these three bands.

3.3. Correlation Analysis between Wave Bands

Figure 3 shows the correlation between wavelength points. Vis-NIR is an indirect technique for rapid measurement and discrimination that requires a small amount of prepared samples and does not use harmful chemicals. It can qualitatively discriminate COD contamination in surface water using spectral absorption characteristics [31]. However, interference of instrument noise and high coincidence of information bands of various components occur during measurement. Vis-NIR also has a wide wavelength range. Therefore, there is extensive irrelevant band information. Figure 3 shows that the correlation between the 1050 wavelength points is different, with some features showing a strong correlation and others showing a weak correlation. Therefore, it is necessary to choose the appropriate wavelength band for modeling and obtain a model with high performance by removing non-informative bands.

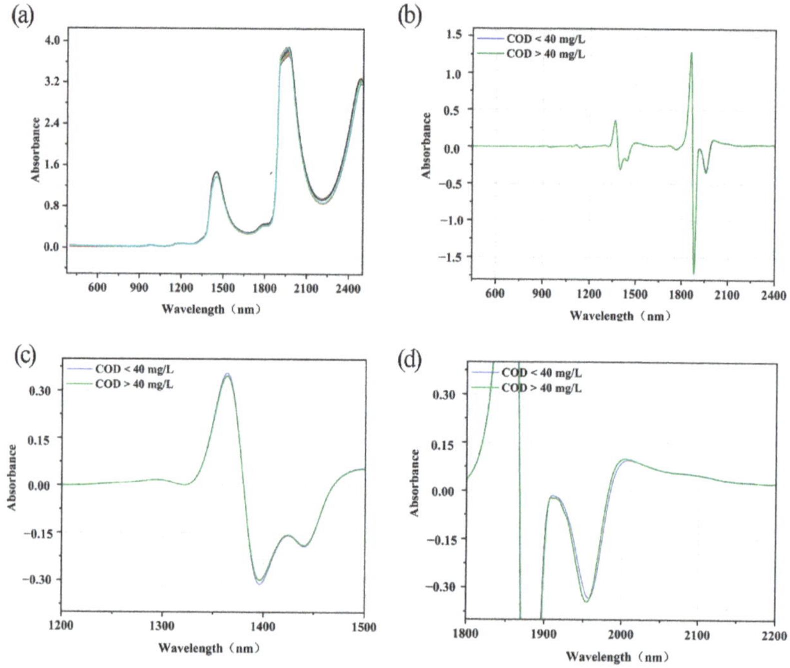

Figure 2. Visible near-infrared spectroscopy plots of surface water samples: (**a**) raw spectra; (**b**) second derivative (SD) (400–2500 nm); (**c**) SD (1200–1500 nm); (**d**) SD (1800–2200 nm).

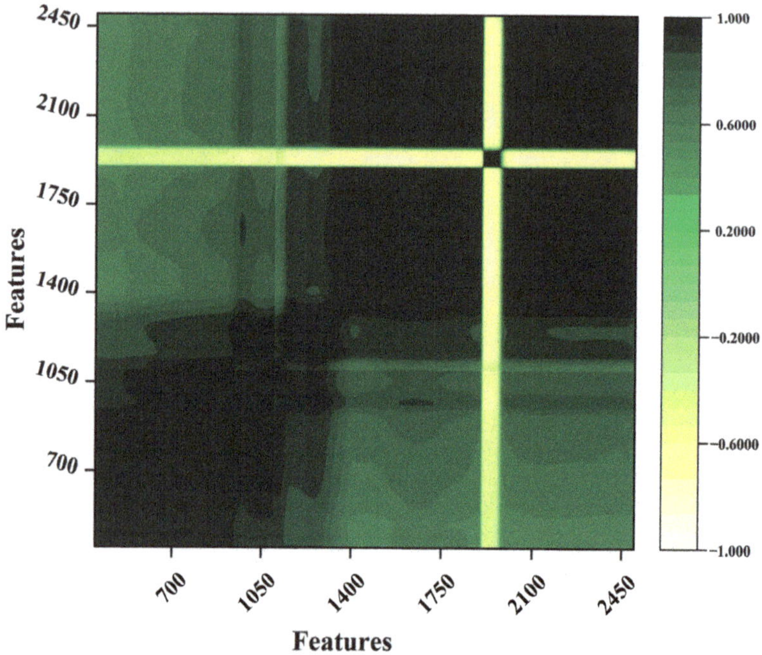

Figure 3. Correlation between 1050 wavelength points.

3.4. Comparison of Classification Results

The discrimination results of the PLS–DA, SMOTE–PLS–DA, and CARS–SMOTE–PLS–DA models with different pre-processing methods for surface water samples are summarized in Table 2. The sample sets were divided using the SPXY method and saved as the training and test sets. The raw spectra were pre-processed differently. In the PLS–DA model, the spectral pre-processing had different effects on surface water pollution discrimination. FD and SD had positive impacts on the accuracy of the model. The model achieved the best prediction results after SD pre-processing. The accuracy of the training set and the accuracy, sensitivity, and specificity of the test set of the PLS–DA model were 0.88, 0.88, 0.83, and 0.93, respectively. However, the MSC and SNV pre-processing methods had a negative impact on the accuracy of the model. With either pre-processing method, the accuracy of the modeling results decreased compared to that with the original spectra. The pre-processed training and test sets were saved separately. The training set was SMOTE-processed to obtain a new training set, which was then subjected to PLS–DA. The SMOTE–PLS–DA modeling results are shown in Table 2. Compared with those of the PLS–DA model, the SMOTE–PLS–DA model accuracy with the FD, SD and MSC pre-processing methods was improved. Among them, for the FD pre-processing method, the training and test set accuracies of the model improved by 7% and 7%. For the SD method, the training and test set accuracies of the model improved by 9% and 6%. For the MSC method, the training and test set accuracies of the model improved by 12% and 3%. However, the accuracy of the SMOTE–PLS–DA model of the SNV pre-processing method was not improved, but the sensitivity of the model was greatly improved.

Table 2. Summary of discrimination results of partial least squares discriminant analysis (PLS–DA), synthetic minority oversampling technique (SMOTE)–PLS–DA, and competitive adaptive reweighted sampling (CARS)–SMOTE–PLS–DA models with different pre-processing methods for surface water.

Sample Type	Model Algorithm	Pre.p *	Number of Bands	Number of Training Sets	Number of Test Sets	Accuracy of Training Sets	Accuracy of Test Sets	Sensitivity	Specificity
Surface water	PLS–DA	RS	1050	95	32	0.85	0.84	0.83	0.86
		FD		95	32	0.87	0.84	0.82	0.90
		SD		**95**	**32**	**0.88**	**0.88**	**0.83**	**0.93**
		MSC		95	32	0.74	0.69	0.68	0.69
		SNV		95	32	0.76	0.75	0.73	0.80
	SMOTE–PLS–DA	RS		140	32	0.89	0.88	0.96	0.63
		FD		140	32	0.94	0.91	0.95	0.90
		SD		**140**	**32**	**0.97**	**0.94**	**0.89**	**0.93**
		MSC		140	32	0.86	0.72	0.63	0.85
		SNV		140	32	0.75	0.75	0.82	0.70
	CARS–SMOTE–PLS–DA	RS	8	140	32	0.88	0.88	0.83	0.93
		FD	10	140	32	0.94	0.94	1.00	0.80
		SD	**38**	**140**	**32**	**0.99**	**0.97**	**0.94**	**1.00**
		MSC	47	140	32	0.83	0.78	0.84	0.69
		SNV	85	140	32	0.85	0.84	0.91	0.70

Notes: * Pre.p: Pre-processing; The boldfaced rows indicate the best pretreatment methods and their results.

To simplify the model and further improve its prediction performance, the raw spectra were pre-processed using four different methods and subjected to feature selection. Then, the training set was processed using the SMOTE algorithm to obtain the results of CARS–SMOTE–PLS–DA, as shown in Table 2. The accuracy of the model improved after CARS feature selection. After SD pre-processing, the training and test set accuracies of the model improved by 11% and 9%, respectively, compared to those of the PLS–DA model. The sensitivity and specificity were greatly enhanced. The simplicity of the model also improved, with 1050 wavelength points being reduced to 38. The CARS algorithm further improved the model performance and simplified the model, compared to the SMOTE–PLS–DA model. The training set accuracy of the model improved and the sensitivity and specificity increased to a greater extent.

To further investigate the performance of the three models, the receiver operating characteristic (ROC) curves of the four different pre-processing methods and surface water score map were plotted and analyzed, the ROC is a comprehensive evaluation index reflecting the continuous variables of the sensitivity and specificity in the classification problem [32], as shown in Figure 4. The points of each curve in Figure 4c are closer to the upper left corner than those in Figure 4a,b, indicating that the prediction accuracy corresponding to each pre-processing method improved with the application of CARS and SMOTE algorithms. However, for the PLS–DA model, the ROC curve is closer to the dashed line after pre-processing with MSC and SNV. In other words, the model performance was reduced. For the SMOTE–PLS–DA model, the ROC curve of the original spectra is closer to the dashed line, i.e., the model performance was poorer. For the CARS–SMOTE–PLS–DA model, compared with Figure 4a, all five curves are closer to the upper left corner, whereas the ROC curves of MSC and SNV are below that of the original spectra. In other words, the MSC and SNV pre-processing methods reduced the model performance. Moreover, the ROC curve of SD is closer to the upper left corner, i.e., the modeling effect with SD pre-processing was better.

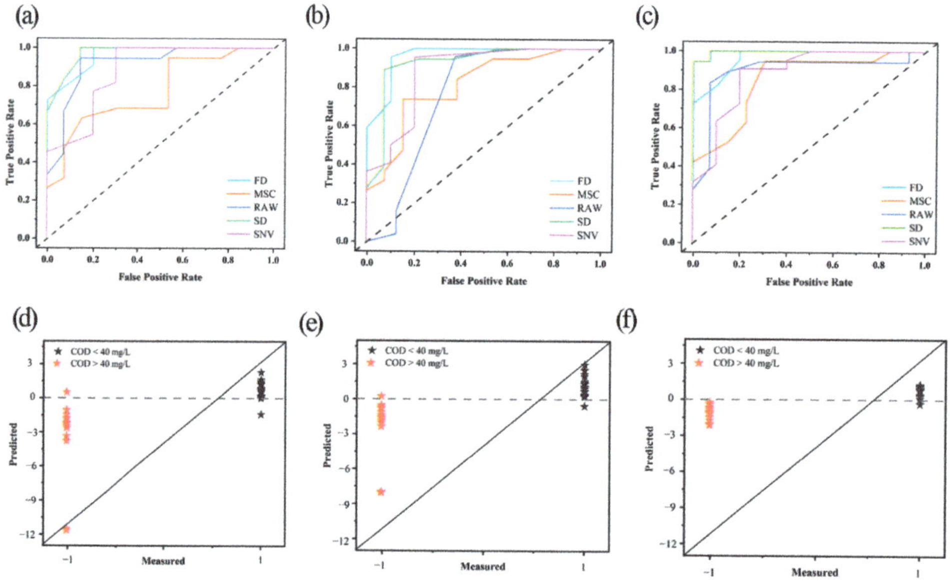

Figure 4. Receiver operating characteristic (ROC) curves and surface water score map: (**a**,**d**) partial least squares discriminant analysis (PLS–DA) model, (**b**,**e**) synthetic minority oversampling technique (SMOTE)–PLS–DA model, (**c**,**f**) competitive adaptive reweighted sampling (CARS)–SMOTE–PLS–DA model. FD: first derivative; SD: second derivative; MSC: multiple scattering correction; SNV: standard normal variate.

In Figure 4d, it can be seen that a large number of samples with a label value of 1 have scores below 0 and a number of samples are misclassified; in Figure 4e, after SMOTE, it can be seen that for samples with the label value of 1, the score has improved significantly, but there are still a number of samples with scores below 0. In order to further improve the score, we used the CARS algorithm to improve the performance of the model. In Figure 4f, we can see that only two samples with the sample label of 1 have scores below 0, at the same time, the scores of the samples with the label −1 are all located below 0, and the model prediction was greatly improved.

4. Discussion

4.1. Band Analysis by CARS Algorithm

The results of the CARS feature selection of the SD pre-processed spectra are shown in Figure 5a. A total of 38 bands were selected as key variables from 1050 wavelength points, mainly located near 430–500, 550–600, 700–860, 1050–1080, 1900–2000, and 2350–2400 nm. To verify whether the selected 38 bands could represent the variability between uncontaminated and contaminated surface water samples, the scores of the bands were plotted, as shown in Figure 5b. There was large variability in the scores of the 38 bands; this also proved the feasibility of these bands selected by CARS. The greatest variability in the scores was found near 498 nm; this may be caused by C-H bond vibrations of aromatic hydrocarbons in the vicinity [28,33].

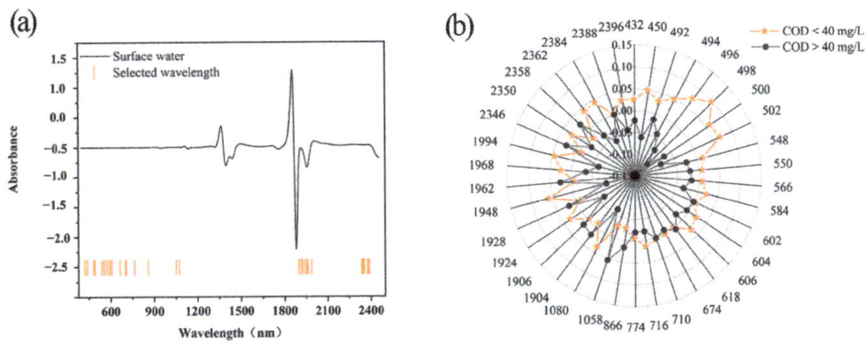

Figure 5. (**a**) Feature bands selected by competitive adaptive reweighted sampling (CARS) after second derivative (SD) pre-processing, (**b**) score plot of the feature bands.

The chemical bonds corresponding to the main bands of the Vis-NIR region screened by CARS and the possible corresponding contamination components are shown in Table 3. The band most screened by CARS was near 400–860 nm; this may arise from the vibration of C-H and N-H chemical bonds, such as those in aromatic hydrocarbons [28,34].

Table 3. Basic chemical bonds, absorption wavelengths, and possible associated water pollution components of main spectral bands screened by competitive adaptive reweighted sampling (CARS) for visible near-infrared region.

Locations of Selected Spectral Variables (nm)	Possible Fundamental Bonds	Possible Related Constituents
800	C-H	Organics (aromatics)
1000	N-H	Organics (amine)
1100	C-H	Organics (aromatics)
1200	C-H	Organics (aromatics)
1380	O-H	Water
1500	C-O	Organics (aromatics)
1800	C-H	Organics
2100	N-H	Organics (amine)
2400	C-O	Organics (Carbohydrates)

4.2. Implication of Proposed Strategy

The CARS–SMOTE–PLS–DA modeling approach proposed in this paper not only improves the discrimination accuracy of the PLS–DA model but also simplifies the model input variables. When using Vis-NIR as the input for the PLS–DA model, most spectral variables may be redundant; on the other hand, fewer spectral input variables may result in the loss of COD-related information. A spectral selection algorithm can solve both problems, and the optimal number of input spectra for a balanced model can be found using spectral variable selection. The modeling effect can reduce due to the large difference between the number of contaminated and uncontaminated surface water samples collected. To solve this problem, the feasibility of the SMOTE algorithm in solving the problem of uneven sample distribution was explored. The feasibility of PLS–DA and SMOTE–PLS–DA was experimentally verified before conducting CARS–SMOTE–PLS–DA. The discrimination accuracy improved after SMOTE solved the problem of the uneven sample distribution. Finally, the Vis-NIR spectra of surface water were subjected to band selection after the pre-processing with four different methods. Combining the CARS selection algorithm with the SMOTE algorithm not only improved the discrimination accuracy of the model but also reduced the input of the discrimination model.

In this study, the surface water samples were collected for a total of 4 months, covering both the rainy and non-rainy seasons in Guangzhou. Changes in the rainy season will lead to changes in COD because the runoff generated by the rainfall in the rainy season will cause pollutants from land sources to enter the water, resulting in an increase in COD. From the principle of COD chemical detection, these pollutants are all aerobic substances. The aerobic substances in the surface water during the rainy season and non-rainy season have a general law and there will be no major changes in components due to the rainy season. We carried out Vis-NIR detection on a large number of samples and used a surface water model to grasp the quantitative relationship between all aerobic substances and COD values as much as possible. We used the CARS–SMOTE–PLS–DA model to realize the online monitoring of large COD values, which provides a new way of discriminating for the management of seriously polluted surface water.

5. Conclusions

This study employed a new approach with CARS–SMOTE–PLS–DA and Vis-NIR to judge whether surface water can meet the COD standard (40 mg/L) for agricultural use and the general landscape. It demonstrated the feasibility and effectiveness of introducing the CARS band selection technique and the SMOTE algorithm into Vis-NIR analysis. The CARS–SMOTE–PLS–DA modeling approach not only had a higher overall accuracy but also produced a more simplified model. The optimal pre-processing method for all three modeling methods was SD, with PLS–DA yielding an accuracy of 88% with the input of 1050 wavelength points. Compared to the PLS–DA model, the CARS–SMOTE–PLS–DA model exhibited an 11% improvement in accuracy and a 96% reduction in wavelength input. The CARS–SMOTE–PLS–DA model experienced a 5% improvement in accuracy and a 96% reduction in wavelength input compared to the SMOTE–PLS–DA model. Overall, the surface water COD discrimination method (CARS–SMOTE–PLS–DA model) proposed in this paper has the advantages of novelty, eco-friendliness, simplicity, and broad prospects. It is a novel method for real-time online surface water COD discrimination, which is conducive to the management and development of surface water resources.

Author Contributions: Conceptualization, X.H., J.M. and W.J.; methodology, X.H., J.M., J.C. and Y.Y.; visualization, X.H., J.C., B.X. and W.Y.; sampling, X.C.; writing—original draft, X.H.; writing—review and editing, X.C., D.X. and F.H. All authors have read and agreed to the published version of the manuscript.

Funding: This work was supported by the National Natural Science Foundation of China (61975069), the Guangzhou science and technology project (202103000095), the Key-Area Research and Development Program of Guangdong Province (2020B090922006), and the Free Exploration Project of Special Research Funds for the Central Public-Interest Scientific Institution (PM-zx703-202112-338).

Institutional Review Board Statement: Not applicable.

Informed Consent Statement: Not applicable.

Data Availability Statement: The data presented in this study are available on request from the corresponding authors. The data are not publicly available due to the continuation of a follow-up study by the authors.

Acknowledgments: The support provided by South China Institute of Environmental Sciences, Ministry of Ecology and Environment, State Environ-ment Protection Key Laboratory of Water Environmental Simulation and Pollution Control, Guangzhou 510655, China.

Conflicts of Interest: The authors declare no conflict of interest.

References

1. Suri, M.R.; Dery, J.L.; Pérodin, J.; Brassill, N.; He, X.; Ammons, S.; Gerdes, M.E.; Rock, C.; Goldstein, R.E.R. U.S. farmers' opinions on the use of nontraditional water sources for agricultural activities. *Environ. Res.* **2019**, *172*, 345–357. [CrossRef] [PubMed]
2. Zhao, Y.; Han, J.; Zhang, B.; Gong, J. Impact of transferred water on the hydrochemistry and water quality of surface water and groundwater in Baiyangdian Lake, North China. *Geosci. Front.* **2021**, *12*, 101086. [CrossRef]
3. Kotir, J.H.; Smith, C.; Brown, G.; Marshall, N.; Johnstone, R. A system dynamics simulation model for sustainable water resources management and agricultural development in the Volta River Basin, Ghana. *Sci. Total Environ.* **2016**, *573*, 444–457. [CrossRef]
4. Shen, M.; Yang, Y. The Water Pollution Policy Regime Shift and Boundary Pollution: Evidence from the change of water pollution levels in C. *Sustainability* **2017**, *9*, 1469. [CrossRef]
5. Ren, S.; Li, Q.; Wang, J.; Fan, B.; Bai, J.; Peng, Y.; Li, S.; Han, D.; Wu, J.; Wang, J.; et al. Development of a fast and ultrasensitive black phosphorus-based colorimetric/photothermal dual-readout immunochromatography for determination of norfloxacin in tap water and river water. *J. Hazard. Mater.* **2021**, *402*, 123781. [CrossRef]
6. Baghanam, A.H.; Nourani, V.; Aslani, H.; Taghipour, H. Spatiotemporal variation of water pollution near landfill site: Application of clustering methods to assess the admissibility of LWPI. *J. Hydrol.* **2020**, *591*, 125581. [CrossRef]
7. Rodrigues, R.P.; Rodrigues, D.P.; Klepacz-Smolka, A.; Martins, R.C.; Quina, M.J. Comparative analysis of methods and models for predicting biochemical methane potential of various organic substrates. *Sci. Total Environ.* **2019**, *649*, 1599–1608. [CrossRef]
8. Rohman, A.; Nugroho, A.; Lukitaningsih, E. Sudjadi Application of vibrational spectroscopy in combination with chemometrics techniques for authentication of herbal medicine. *Appl. Spectrosc. Rev.* **2014**, *49*, 603–613. [CrossRef]
9. Yuan, H.; Liu, C.; Wang, H.; Wang, L.; Dai, L. PLS-DA and Vis-NIR spectroscopy based discrimination of abdominal tissues of female rabbits. *Spectrochim. Acta A* **2022**, *271*, 120887. [CrossRef]
10. Zhang, M.; Zhang, B.; Li, H.; Shen, M.; Tian, S.; Zhang, H.; Ren, X.; Xing, L.; Zhao, J. Determination of bagged 'Fuji' apple maturity by visible and near-infrared spectroscopy combined with a machine learning algorithm. *Infrared Phys. Technol.* **2020**, *111*, 103529. [CrossRef]
11. Xie, X.-L.; Li, A.-B. Identification of soil profile classes using depth-weighted visible–near-infrared spectral reflectance. *Geoderma* **2018**, *325*, 90–101. [CrossRef]
12. Begum, N.; Maiti, A.; Chakravarty, D.; Das, B.S. Diffuse reflectance spectroscopy based rapid coal rank estimation: A machine learning enabled framework. *Spectrochim. Acta A* **2021**, *263*, 120150. [CrossRef] [PubMed]
13. Sharififar, A.; Sarmadian, F.; Minasny, B. Mapping imbalanced soil classes using Markov chain random fields models treated with data resampling technique. *Comput. Electron. Agric.* **2019**, *159*, 110–118. [CrossRef]
14. Vohland, M.; Ludwig, M.; Thiele-Bruhn, S.; Ludwig, B. Determination of soil properties with visible to near- and mid-infrared spectroscopy: Effects of spectral variable selection. *Geoderma* **2014**, *223–225*, 88–96. [CrossRef]
15. Yu, K.; Fang, S.; Zhao, Y. Heavy metal Hg stress detection in tobacco plant using hyperspectral sensing and data-driven machine learning methods. *Spectrochim. Acta A* **2021**, *245*, 118917. [CrossRef]
16. Han, X.; Xie, D.; Song, H.; Ma, J.; Zhou, Y.; Chen, J.; Yang, Y.; Huang, F. Estimation of chemical oxygen demand in different water systems by near-infrared spectroscopy. *Ecotox. Environ. Saf.* **2022**, *243*, 113964. [CrossRef]
17. Ma, Y.; Tie, Z.; Zhou, M.; Wang, N.; Cao, X.; Xie, Y. Accurate determination of low-level chemical oxygen demand using a multistep chemical oxidation digestion process for treating drinking water samples. *Anal. Methods* **2016**, *8*, 3839–3846. [CrossRef]
18. *GB 3838-2002*; Environmental Quality Standards for Surface Water. Ministry of Ecology and Environment of China (CNMEE): Beijing, China, 2002. (In Chinese)
19. Hong, Y.; Liu, Y.; Chen, Y.; Liu, Y.; Yu, L.; Liu, Y.; Cheng, H. Application of fractional-order derivative in the quantitative estimation of soil organic matter content through visible and near-infrared spectroscopy. *Geoderma* **2019**, *337*, 758–769. [CrossRef]
20. Galvao, R.K.; Araujo, M.C.; Jose, G.E.; Pontes, M.J.; Silva, E.C.; Saldanha, T.C. A method for calibration and validation subset partitioning. *Talanta* **2005**, *67*, 736–740. [CrossRef]
21. Xu, Y.; Goodacre, R. On splitting training and validation Set: A comparative study of cross-validation, bootstrap and systematic sampling for estimating the generalization performance of supervised learning. *J. Anal. Test.* **2018**, *2*, 249–262. [CrossRef]
22. Sharififar, A.; Sarmadian, F.; Malone, B.P.; Minasny, B. Addressing the issue of digital mapping of soil classes with imbalanced class observations. *Geoderma* **2019**, *350*, 84–92. [CrossRef]
23. Chawla, N.V.; Bowyer, K.W.; Hall, L.O.; Kegelmeyer, W.P. SMOTE: Synthetic minority over-sampling technique. *J. Artif. Intell. Res.* **2002**, *16*, 321–357. [CrossRef]
24. Yang, X.; Li, Y.; Wang, L.; Li, L.; Guo, L.; Huang, F.; Zhao, H. Determination of 10-Hydroxy-2-Decenoic acid of royal jelly using near-infrared spectroscopy vombined with chemometrics. *J. Food Sci.* **2019**, *84*, 2458–2466. [CrossRef] [PubMed]
25. Li, H.; Liang, Y.; Xu, Q.; Cao, D. Key wavelengths screening using competitive adaptive reweighted sampling method for multivariate calibration. *Anal. Chim. Acta* **2009**, *648*, 77–84. [CrossRef] [PubMed]

26. Luo, N.; Yang, X.; Sun, C.; Xing, B.; Han, J.; Zhao, C. Visualization of vibrational spectroscopy for agro-food samples using t-Distributed Stochastic Neighbor Embedding. *Food Control* **2021**, *126*, 107812. [CrossRef]
27. Zhou, Q.; Huang, W.; Fan, S.; Zhao, F.; Liang, D.; Tian, X. Non-destructive discrimination of the variety of sweet maize seeds based on hyperspectral image coupled with wavelength selection algorithm. *Infrared Phys. Technol.* **2020**, *109*, 103418. [CrossRef]
28. Cen, H.; He, Y. Theory and application of near infrared reflectance spectroscopy in determination of food quality. *Trends Food Sci. Tech.* **2007**, *18*, 72–83. [CrossRef]
29. Rossel, R.A.V.; Behrens, T. Using data mining to model and interpret soil diffuse reflectance spectra. *Geoderma* **2010**, *158*, 46–54. [CrossRef]
30. Xu, L.; Hong, Y.; Wei, Y.; Guo, L.; Shi, T.; Liu, Y.; Jiang, Q.; Fei, T.; Liu, Y.; Mouazen, A.M.; et al. Estimation of organic carbon in anthropogenic soil by VIS-NIR spectroscopy: Effect of variable selection. *Remote Sens.* **2020**, *12*, 3394. [CrossRef]
31. Andreu-Rodríguez, J.; Pérez-Espinosa, A.; Pérez-Murcia, M.D.; Moral, R.; Agulló, E.; Ferrández-Villena, M.; Ferrández-García, M.T.; Bustamante, M.A. Near infrared reflectance spectroscopy (NIRS) for the assessment of biomass production and C sequestration by arundo donax L. in salt-affected environments. *Agric. Water Manag.* **2017**, *183*, 94–100. [CrossRef]
32. Daniel, M.; Guerra, R.; Brázio, A.; Rodrigues, D.; Cavaco, A.M.; Antunes, M.D.; Valente de Oliveira, J. Feature discovery in NIR spectroscopy based Rocha pear classification. *Expert Syst. Appl.* **2021**, *177*, 114949. [CrossRef]
33. Hong, Y.; Chen, Y.; Shen, R.; Chen, S.; Xu, G.; Cheng, H.; Guo, L.; Wei, Z.; Yang, J.; Liu, Y.; et al. Diagnosis of cadmium contamination in urban and suburban soils using visible-to-near-infrared spectroscopy. *Environ. Pollut.* **2021**, *291*, 118128. [CrossRef] [PubMed]
34. Cozzolino, D.; Smyth, H.E.; Gishen, M. Feasibility study on the use of visible and near-infrared spectroscopy together with chemometrics to discriminate between commercial white wines of different varietal origins. *J. Agric. Food Chem.* **2003**, *51*, 7703–7708. [CrossRef] [PubMed]

water

Article

Marine Environmental Capacity in Sanmen Bay, China

Yanming Yao [1], Jiahao Zhu [1], Li Li [1,2,*], Jiachen Wang [1] and Jinxiong Yuan [3]

1 Ocean College, Zhejiang University, Zhoushan 316021, China; hotfireyao@163.com (Y.Y.); 21934102@zju.edu.cn (J.Z.); 22034207@zju.edu.cn (J.W.)
2 The Engineering Research Center of Oceanic Sensing Technology and Equipment, Ministry of Education, Zhoushan 316021, China
3 Hangzhou Xi'ao Environmental Technology Co., Ltd., Hangzhou 310005, China; yjx3051243007@163.com
* Correspondence: lilizju@zju.edu.cn

Abstract: Estuarine environmental capacity is the foundation for coastal biological diversity and self-purification capacity. Hence, studies on the marine environmental capacity (MEC) are the foundation for the total discharge control and water quality improvement of land-based pollutants. In the article, A calibrated two-dimensional hydrodynamic model was used to study the environmental characteristics of Sanmen Bay, including the tides, the residual currents, the tidal prism, and water exchange abilities. The model results were used to estimate the environmental capacity of the bay. Taking the pollution problem in Sanmen Bay as an example, the method of response factor, the sub-unit control method, and the phased control method were used to estimate the environmental capacity, pollutant amounts, and the pollutant reduction in the bay. The concentrations of COD, inorganic nitrogen, and acid salt in Sanmen Bay are spatially varied, with higher values occurring in the western part and in the inner bay. The half exchange time of the whole bay is about 23 days, and the exchange time of 95% water body is about 60 days. The evaluation of MEC cannot only provide technical support for the offshore aquaculture industries but also provide a scientific basis for the total control of terrigenous pollutants in coastal cities in Southern Zhejiang Province.

Keywords: marine environmental capacity (MEC); response factor; water exchange; residual; Sanmen Bay

Citation: Yao, Y.; Zhu, J.; Li, L.; Wang, J.; Yuan, J. Marine Environmental Capacity in Sanmen Bay, China. *Water* **2022**, *14*, 2083. https://doi.org/10.3390/w14132083

Academic Editor: Karl-Erich Lindenschmidt

Received: 24 May 2022
Accepted: 23 June 2022
Published: 29 June 2022

1. Introduction

With the fast development of the coastal economy and marine aquaculture, estuaries are suffering a lot from terrestrial and oceanic pollutants. Understanding the characteristics of pollutants and estimating the marine environmental capacity (MEC) in estuaries provides a theoretical foundation for coastal management. The deteriorated environment of bays feeds back to and constrains the economy of coastal cities. Therefore, identifying sources and sinks of pollutants, and estimating the estuarine environmental capacity is quite important for coastal development (Yoon et al., 2020; Halpern et al., 2008; Syvitski et al., 2005) [1–3].

The main reason for the deterioration of the offshore water environment is that a large number of pollutants produced on land are discharged into it, including inorganic nitrogen, active phosphate, and heavy metals, etc. (Chen et al., 2008; Cui et al., 2013) [4,5].

The first step to conduct total discharge control is to assess the MEC of pollutants. (Wu et al., 2005) [6]. MEC refers to the maximum load of pollutants that can be accommodated in a specific sea area under the premise of making full use of the marine self-purification capacity without causing pollution damage (Linker et al., 2013) [7].

Sanmen Bay is located on the coast of the East China Sea (Figure 1a). It is a macro-tidal turbid estuary, with a maximum tidal range of approximately 2 m at the bay mouth and suspended sediment concentration (SSC) of 1.192 kg/m^3 at the middle of the bay. It has four tributaries. With the fast development of human activities, the pollution problem is becoming increasingly heavy in the bay.

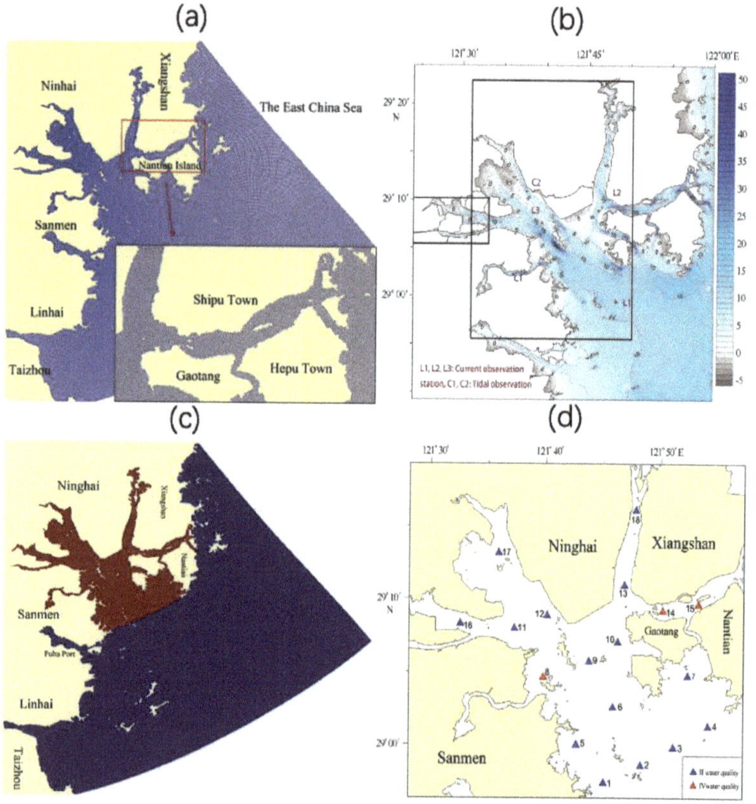

Figure 1. (**a**) Model domain for Sanmen Bay, (**b**) Location of observation stations, (**c**) Initial conditions of water exchange model (The concentration distribution in Sanmen Bay is set as 1 unit, red in the figure, the inflow outside the bay and at the boundary is set as 0, dark blue in the figure), (**d**) Water quality control stations in Sanmen Bay.

Identification of the contamination source is a part of the systematical analysis to find the pollution source fundamentally. Because how much a certain water body can hold is on the premise of pollution intensity which also plays the role to decide how to distribute for such a specific object.

According to the various forms of pollution being discharged into the water body, pollution sources can be divided into point source pollution and non-point source pollution. Non-point source pollution is referred to as water body pollution caused by rainfall run-off, this kind of contaminant entered into the soil or underground water body in a wide, microcrystalline, and dispersive way. Since the 1960s, research on non-point source pollution has caught the interest of scientists throughout the world. So far non-point source pollution feature, impact factor, load quantification on pollution output, and mechanism of pollutant migration and transformation have made great achievements.

A certain amount of contaminant emission poured is permitted. Because natural water body holds a certain amount of environmental capacity for a sort of pollution. Total emission radically depends on assimilative capacity, distribution of total water pollutants based on the limit of water environmental capacity.

According to the environmental quality survey results of the Sanmen Bay sea area from 2015 to 2016, and combined with the marine environmental function zoning, the researchers evaluated and classified the current situation of the environmental quality of the sea area (Liang et al., 2021) [8]. A 2-D hydrodynamic and pollutant model of Sanmen Bay is established based on Deift3D. Combined with the current hydrological status of Sanmen Bay, the transport and diffusion laws of COD, TP, and TN in the bay are analyzed (He et al., 2018) [9].

After conducting Clean Water Act in 1972, although industrial and municipal pollution has been under effective control by carrying out NPDES (National Pollutant Discharge Elimination System), the water quality has not been radically improved. Previous research showed that it is non-point source pollution that mainly polluted the rivers, lakes, and surface waters in estuaries. Also, non-point source pollution polluted underground water and degradation of wetland ecosystem. For such a reason, the Clean Water Act includes a guide rule, named TMDL (Total Maximum Daily Load), aiming at controlling both point source pollution and non-point source pollution. The acting emphasis is on figuring out non-point source pollution load and elimination in key water areas. Such a concept (MEC) was raised by Japanese scholars in the environment field in 1968. Japanese researcher Yanowokio (1968) claimed that environmental capacity is determined by environment quality standards, i.e., keeping total contamination loads within a permitted limit [10].

Streeter and Phelps (1952) raised a simple S-P model, which is the earliest form of water quality model [11]. The development of the water quality model can be concluded into 5 periods: 1925–1960, BOD-DO coupled model was raised based on the S-P model; during 1960–1965, spatial variety, physical, kinetic factors, and temperature were introduced into the model as a state variable, in the same the heat exchange between air and the water surface was also considered; during 1965–1970, as the computer started to be applied, people increasingly deeply understand biochemical oxygen consumption; the calculation method was developed from 1-D to 2-D; during 1970–1975, water quality model has developed into mutually non-linearity. In the last 20 years, space dimensionality has been into 3-D, the emphasis of study gradually turned into improving the dependability and evaluation of the model. In this way, dependability on the calculation of environmental capacity has been enhanced [12–16].

The eutrophication degree of seawater, the enrichment degree of heavy metals in sediments and the potential ecological hazard effects were comprehensively analyzed.

In this study, we take Sanmen Bay as an example, to study the characteristics of water exchangeability and marine environmental capacity, using both numerical models and field data.

Sanmen Bay is located in the monsoon subtropical humid climate zone. Affected by the monsoon climate, it has four distinct seasons and a mild climate. The weather changes are complex and disastrous weather is frequent. Disastrous weather of different degrees can be encountered in all seasons. The seasonal variation of temperature is obvious. Sanmen Bay has abundant rainfall, mainly from March to September. The whole year can be roughly divided into two rainy seasons and two relatively dry seasons. The wind direction of Sanmen Bay varies with seasons. Typhoons, rainstorms, and sudden small-scale disastrous weather occur from time to time.

2. Materials and Methods

2.1. Model Descriptions

A calibrated two-dimensional hydrodynamic model was used to reproduce the coastal oceanic and environmental characteristics of Sanmen Bay. Delft 3D simulates water surface elevation, velocity, water quality, waves, and morphology. Flow, Hydrodynamics were used in this study. The governing equations of the Delft3D hydrodynamic model in the vertical sigma coordinate system and horizontal curvilinear coordinate system are expressed as follows:

$$\frac{\partial \zeta}{\partial t} + \frac{1}{\sqrt{G_{\xi\xi}}\sqrt{G_{\eta\eta}}} \frac{\partial\left((d+\zeta)U\sqrt{G_{\eta\eta}}\right)}{\partial \xi} + \frac{1}{\sqrt{G_{\xi\xi}}\sqrt{G_{\eta\eta}}} \frac{\partial\left((d+\zeta)V\sqrt{G_{\xi\xi}}\right)}{\partial \eta} = (d+\zeta)Q \quad (1)$$

where ξ is the coordinate direction under the Delft3D curvilinear coordinate system corresponds to the X-axis of the rectangular coordinate system, η is the Y-axis, ζ is the height of the water surface above the zero scale line of the Z coordinate, d is the depth from the zero scale line of the Z coordinate to the bottom of the water, U is the velocity for X-axis, V is the velocity for Y-axis, $\sqrt{G_{\eta\eta}}$ is conversion coefficient for X-axis and $\sqrt{G_{\xi\xi}}$ is conversion coefficient for Y-axis.

The three-dimensional convection-diffusion equation in the water quality module is as follows:

$$\frac{Cover\partial C}{\partial t} + v_x\frac{\partial C}{\partial x} - D_x\frac{\partial^2 C}{\partial x^2} + v_y\frac{\partial C}{\partial y} - D_y\frac{\partial^2 C}{\partial y^2} + v_z\frac{\partial C}{\partial z} - D_z\frac{\partial^2 C}{\partial z^2} = S + f_R(C,t) \quad (2)$$

where C is substance concentration, D is diffusion coefficient, S is the inflow term, $f_R(C,t)$ is the reaction term.

2.2. Model Configurations

The model domain contains Sanmen Bay and its adjacent seas. The open ocean boundary is from 28°31′ N to 29°26′ N, the east part can extend to 122°27′ E. Jiaojiang River runoff is considered at the open boundary. The measurements were conducted in the bay between the Nantian station and the Linhai station (Figure 2a).

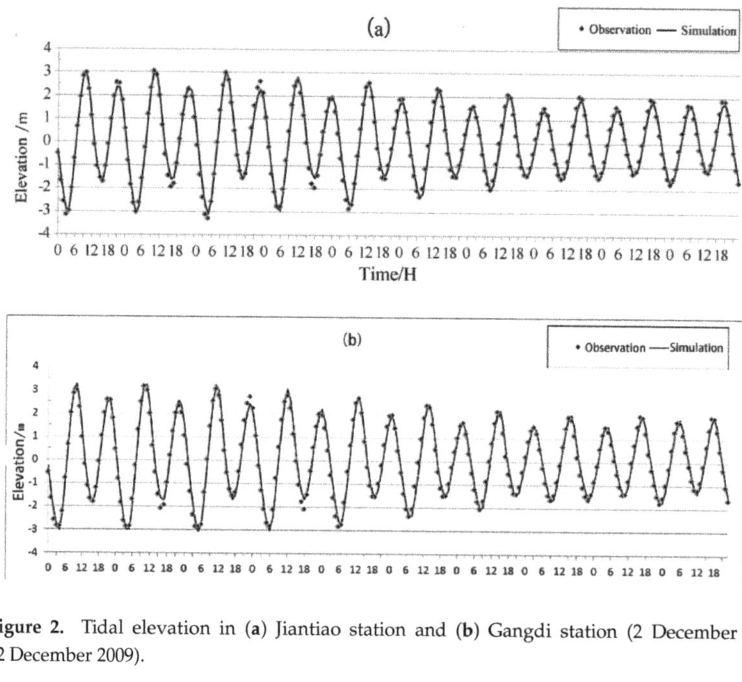

Figure 2. Tidal elevation in (**a**) Jiantiao station and (**b**) Gangdi station (2 December 2009 to 12 December 2009).

According to the calculation region, to generate orthogonal curvilinear grid scatter with the 2-D hydrodynamic model. The Mesh quantity is 629 × 674. In the study region, the minimal edge length is about 50 m, the maximum edge length is about 220 m, and the computational time step is set at 60 s. Charted depth data is gained from the historical chart. ADI method is used to solve the problem.

The computational domain is about 100 km on the x-axis and 110 km on the y-axis. The grid consisted of 423,946 elements, forming a mesh of orthogonal curvilinear grid with variable cell widths ranging from 220 m in the area of the open sea to 50 m in Sanmen Bay. The bathymetry data were interpolated linearly and were corrected with the satellite chart (Figure 1a).

The model was run for a further one month for the period from 1 December 2009 to 30 December 2009.

To calculate the environment capacity, we firstly determine water quality goal by dividing functional the water area, then conduct a numerical simulation to consider the quantitative response relation between pollution emissions and water environmental quality.

2.3. Model Validation

Hourly water elevation from 3 December to 12 December in 2009 is taken from Jiantiao station and Gangdi station are compared with model elevations in Figure 2. The result of verification basically agreed with the measured data. The relative error of the whole process could be controlled within 10%. According to spring tides and neap tides verification, the result can basically reflect the tide wave transformation of Sanmen Bay.

The current velocity and direction of each verification point coincides with the measured data, what's more, the current velocity summit and direction changing moment are both close to each other, and relative error could be controlled under 20%. In a conclusion, the numerical simulation in Sanmen Bay can basically reflect the regional hydrodynamic situation, which offers a basis for further research on water exchange and water environmental capacity (Figures 3 and 4).

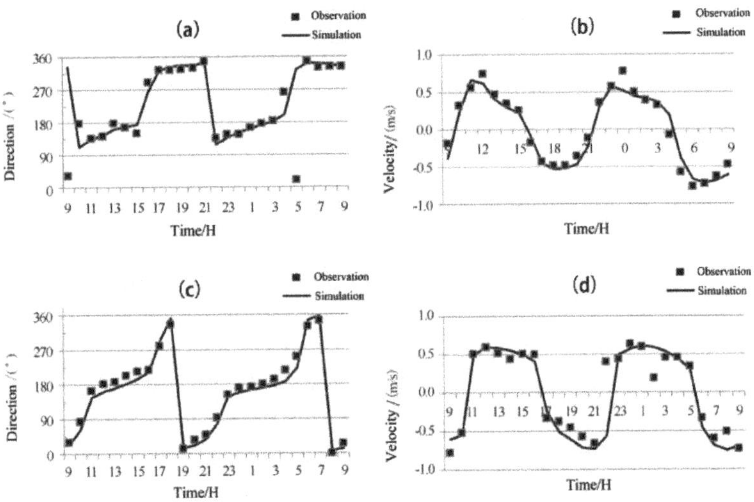

Figure 3. Current velocity and direction during spring tides (**a**,**c**) stand for direction, (**b**,**d**) stand for velocity.

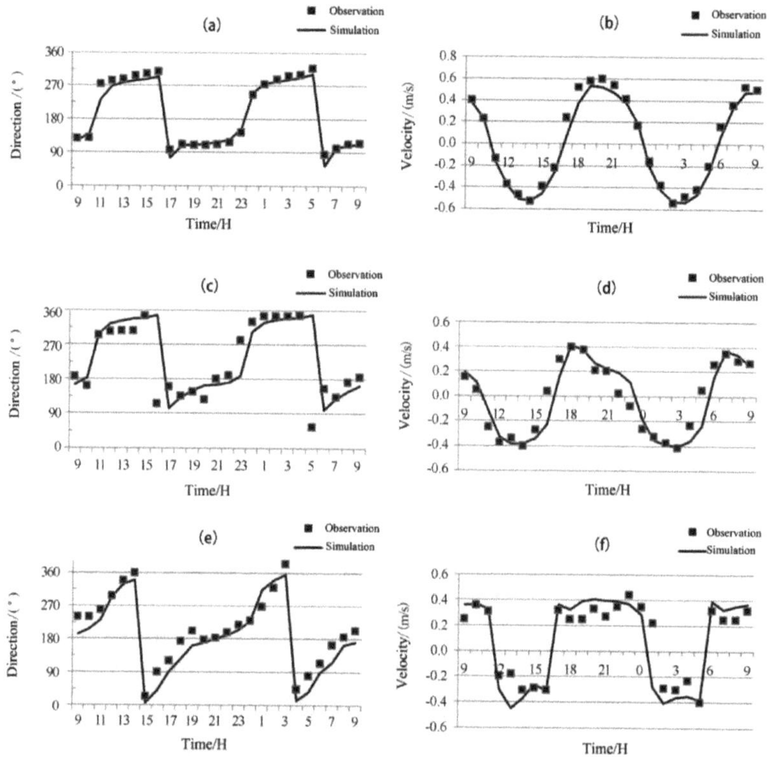

Figure 4. Current velocities and direction during neap tides, (**a**,**c**,**e**) stand for direction, (**b**,**d**,**f**) stand for velocity.

Compared with the measured values, the calculated values of the total tidal mean concentration in the whole sea area are also close, with an error of only 1.96%, indicating that the water quality model has successfully simulated the distribution of inorganic nitrogen concentration in Sanmen Bay.

The validation of COD is shown in Table 1. The model can simulate the contaminants well.

Table 1. Validation of COD.

Station	Flood (Model)	Ebb (Model)	Average (Model)	Observation	Error/(%)
S1	0.6777	0.7055	0.6916	0.717	−3.60
S2	0.6225	0.6749	0.6487	0.661	−1.91
S3	0.5673	0.6360	0.6016	0.640	−5.93
S4	0.5399	0.5942	0.5671	0.556	2.06
S5	0.5376	0.5483	0.5430	0.535	1.49
S6	0.6963	0.7078	0.7020	0.728	−3.51
S7	0.6539	0.6997	0.6768	0.630	7.43
S8	0.6105	0.6739	0.6422	0.583	10.25
S9	0.6986	0.7020	0.7003	0.660	6.06
S10	0.6865	0.6965	0.6915	0.594	16.51
Average	0.6432	0.6658	0.6545	0.642	1.96

2.4. Method to Estimate the MEC

The marine environment capacity (MEC) is decided by the following factors:

(1) The hydrogeology condition of the marine environment, such as marine space, locations, tidal conditions, self-purification abilities, and other natural conditions as well as group features of the ocean ecosystem.
(2) Rules on the utilization function of the specific maritime area usually differ. Different maritime functional areas carry out different water quality standards so that their water capacity varies.
(3) Physicochemical property. The environment capacity relies on self-purification which is decided by the physicochemical property of contaminants. Various kinds of contaminations do harm to people through different toxicities. Thus, environment capacity changes as the concentration permitted to exist differ.

The response coefficient method is adopted here for the research on environment capacity. By taking residual capacity maximum as principal, this study has Sanmen Bay for example, and then calculates the distribution of the marine environment capacity of Sanmen Bay.

Given that velocity and diffusion coefficient is determined, a convection-diffusion equation can be regarded as a linear equation.

The formula is shown as follows:

$$C(x,y,z) = \sum_{i=1}^{m} C_i(x,y,z) \tag{3}$$

where $C(x,y,z)$ means the concentration of each position (mg/L). $C_i(x,y,z)$ means the concentration of the ith position (mg/L).

The concentration field formed separately by each source can also be regarded as some times of the one formed when a certain source strongly discharges pollution.

$$C_i(x,y,z) = Q_i \cdot \alpha_i(x,y,z) \tag{4}$$

where Q_i is the ith pollution source's emission; α_i is the response coefficient field of the ith pollution source, referring to one point's concentration under unit source intensity. α_i reflects the ith point's response level to the ith pollution source.

According to the response coefficient and controlling goals of each controlling point, to work out the environment capacity by the linear programming method. The main steps are as follows:

(1) After working out each source's response coefficient field, researchers should extract each controlling point's corresponding response coefficient value.
(2) Pollution source's permitted emission might correspond to the total residual emission maximum worked out by linear programming (LP) under the condition where controlling points meet the demand of water quality.

Taking Sanmen Bay, for example, this paper studies total amount control and emission cutdown management. One of the central basis of management of the marine environment is environment capacity. Contaminants picked to be calculated should reflect the water quality, degree of contamination, and operability on the management of environment capacity and contamination controlling and other aspects. Considering pollution source, the water quality of Sanmen Bay, and linking to the control index of land-sourced pollutants, Sanmen Bay's prime pollutants are nutrient salt formation such as nitrogen, phosphorus, etc. COD is a comprehensive index to describe the degree of water pollution. According to relevant provisions, CODs, (chemical oxygen demand) TN (total nitrogen), and TP (total phosphorus) should be taken into consideration in calculation of environment capacity or cutdown. Based on investigations of Sanmen Bay and relevant provisions on seawater quality, this paper will calculate by controlling the content of COD_{Mn}, inorganic nitrogen and labile phosphate in seawater and then gets a result. Finally, the research will make a conversion from the result to environment capacity or cutdown for COD, TN, and TP.

3. Results

3.1. Hydrodynamics in the Bay

Field Observation and Analysis

Surface water elevations were measured by WSH at the Jiantiao station and Gangdi station, both recorded at a 60-min sampling interval. The velocity data discussed in this paper are measured by ADP at 400 KHz and 600 KHz in these two stations.

The tidal characteristic values of Jiantiao Station and Gangdi Station are less than 0.5, so it can be inferred that the tide of Sanmen Bay belongs to the regular semidiurnal tide (Table 2).

Table 2. Tidal eigenvalue (H_{O_1}, H_{K_1} and H_{M_2} means the amplitudes of H_1, O_1 and M_2 respectively).

Date	$(H_{O_1}+H_{K_1})/H_{M_2}$	
	Jiaotiao Station	Gangdi Station
2009.7	0.36	0.32
2009.12	0.36	0.33

The current property ratios $((W_{O1} + W_{K1})/W_{M2})$ of L1, L2, and L3 are all less than 0.5, indicating that the M_2 tidal component of Sanmen Bay is dominant and is the main tidal component. The tidal current type of Sanmen Bay belongs to the regular half day tidal current. Generally, the bay is greatly affected by the tidal component of the shallow sea. The calculation results also prove this point. Among them, the value (W_{M4}/W_{M2}) of theL1 station is the smallest. This is mainly because the L1 station is located at the mouth of Sanmen Bay, with an open water surface and deep depth, so it is less affected by the tidal component of the shallow sea (Table 3).

Table 3. Tidal current eigenvalue (W_{O_1}, W_{K_1} and W_{M_1} means the length of long semi axis of partial current ellipse of O_1, K_1 and M_1 respectively).

Station	Layer	$(W_{O_1}+W_{K_1})/W_{M_2}$		W_{M_4}/W_{M_2}	
		July 2009	December 2009	July 2009	December 2009
L1	surface	-	0.283	-	0.059
	0.2H	-	0.227	-	0.209
	0.6H	-	0.244	-	0.049
	0.8H	-	0.235	-	0.147
	bottom	-	0.267	-	0.133
L2	layer	0.104	0.222	0.213	0.250
	bottom	0.061	0.292	0.182	0.208
L3	layer	0.235	0.211	0.250	0.452
	bottom	0.206	0.348	0.267	0.688

3.2. Model Result Hydrodynamics in the Bay

Figure 5 shows the torrent velocity vector of ebb and flood tide during spring and neap tide in December 2009.

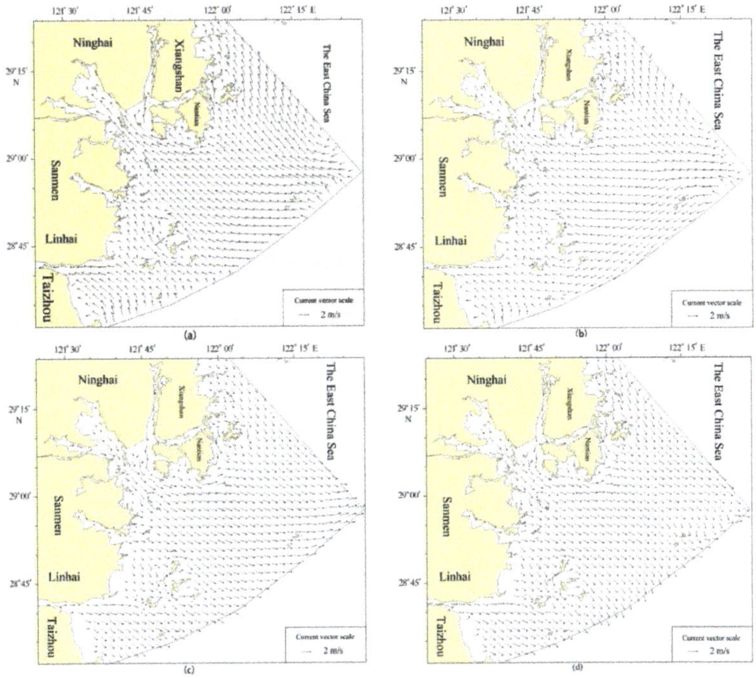

Figure 5. Depth-average velocity vectors during flood and neap (the entire domain). ((**a**,**b**) flood spring and flood ebb, (**c**,**d**) neap spring and neap ebb).

After the tidal waves propogate from the outer sea into Sanmen Bay through southeast to the northwest, the tidal motion characteristics are formed mainly with standing waves in the bay, while ebb and flood last little. The tidal current in Sanmen Bay is basically reciprocating, and the flow direction is mostly influenced by the topography. The flood tide flowing into the bay mouth is mainly in the northwest direction, and the ebb tide is mainly in the southeast direction. The water channels and harbor branches are basically along the longitudinal axis, among them Shipu Harbor's flood tide flow to the west, and ebb tide flow to the east. The open sea area from the bay mouth to the outer bay shows swirling current characteristics in different degrees.

During the flood period, the East China Sea tide waves propagate from the southeast side of the large area along the northwest direction, which mainly enters Sanmen Bay through its mouth. Most of the flood tide flows into the bay in the northwest through the bay mouth after passing through the Maotou Sea. A small part of the tide flows into the bay from east to west via Shipu Port. The two flood tides merge in the bay and then divide into four flows towards the summit of the bay. The first one flows into Jiantiao Harbor, the second one flows into Qimen Harbor and Haiyou Harbor through the Shepan water channel, the third one flows into Liyang Harbor and Qingshan Harbor, the fourth one flows into Baijiao water channel. The direction of the flood tide in each port branch is basically parallel to the longitudinal axis of the port branch, and the velocity is similar to that of the flood flow in the open sea. On the large area of shallow shoals which are apart from the port branch, the floodplain shows a slow flow and diffusing state.

During the ebb flow, the flow direction is basically opposite to that of flood flow, and the main stream is along southeast. The water at the bay summit leaks out of the trough. After the ebb tides converge, most of them flow out of the Sanmen Bay mouth in the southeast direction, and a small part of it flows out from west to east via Shipu Port. The Yushan Islands which is in the southeast corner of the calculation area and the Dongji

Islands in the southwestern part are both affected by the underwater topography, and the direction of the flood and ebb tides change to a certain extent, and the local currents are relatively complicated.

3.3. Water Exchange Ability in the Bay

Luff et al. (1996) introduced the concept of the Half-life period, which is defined as the time required for the concentration of the conservative substance to be diluted to half of the initial concentration by convection diffusion. The definition is based on the fact that it is almost impossible for the final concentration of a substance to be zero, and the rate of dilution represents the rate of water quality change, that is, the exchange capacity of the sea area.

Based on the concept of half exchange time, this study calculated the diffusion, transport, and dilution speed of conservative matter in each grid point of Sanmen Bay by using the transport and diffusion model of conservative materials, so as to study the water exchange capacity of Sanmen Bay.

3.3.1. Passive-Tracer Concentrations

Based on the previous hydrodynamic model, the water exchange model for the transport of regional passive-matter concentration is established (Figure 1c).

In the closed boundary condition, the current is 0 ($\frac{\partial C}{\partial n} = 0$), different treatment methods are adopted for the inflow and outflow of the model, and the inflow material concentration is 0, and the outflow material concentration is calculated by the model. The flow conditions are automatically obtained from the flow model.

According to the conservative material model, the whole continuous tidal process is applied as the calculation tide pattern, and the diffusion, transportation, and dilution process of the conservative material in the calculation water area are obtained continuously. Figure 6 shows the distribution of conservative materials at different times over three months. Note that the concentration value of conservative substances on each grid point not only represents its own concentration, but also is an important indicator of local water exchange degree. The period of the water exchange rate reaching 95% is about 60 days.

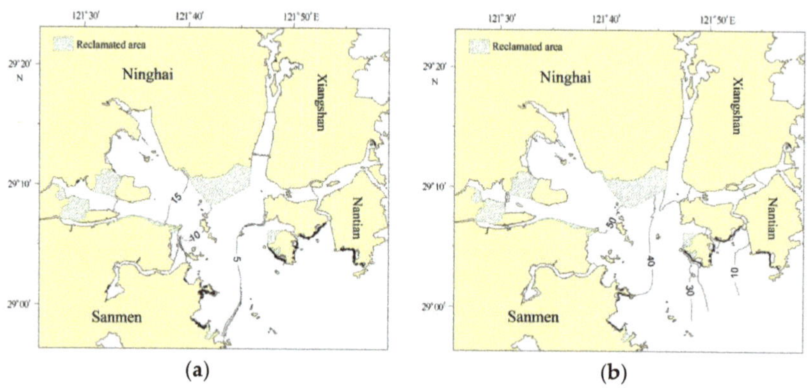

Figure 6. Water semi-exchange time (**a**) and 95% water exchange time (**b**).

After five days (Figure 7a), the water exchange degree of different regions in the bay is quite different. On the whole, the concentration of conservative material decreased gradually from the top of the bay to the estuary, indicating an increased exchange degree from the bay head to the bay mouth. At the same time, the water exchange degree of the bay mouth section increased gradually from the west to the east part of the bay. The direction of concentration isoline near the bay mouth was NNE-SSW, and the water exchange rate reached more than 90% on the fifth day in the bay mouth. The water exchange capacity of

Shipu port is relatively strong, and the water exchange rate from the bay head to the bay mouth rises from 40% to about 80%.

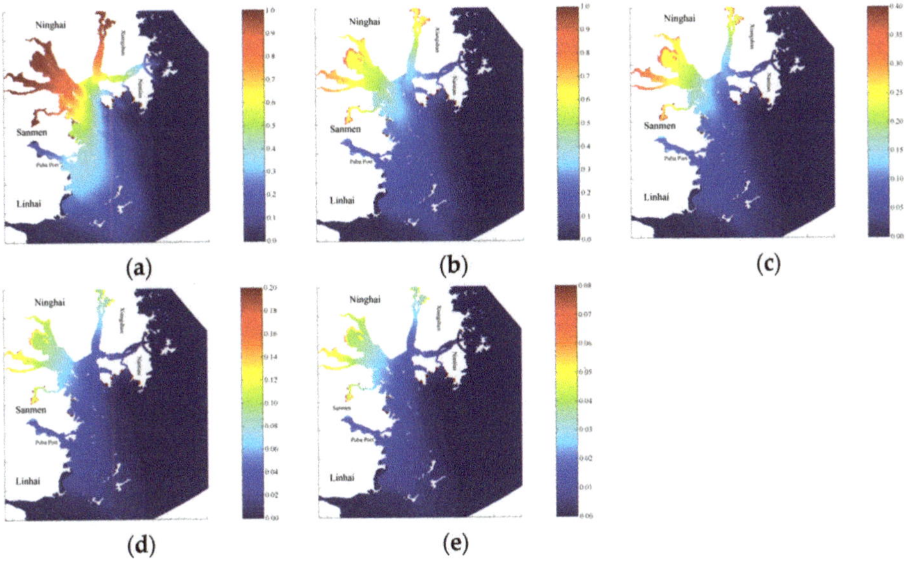

Figure 7. Conserved matter distribution after 5 days (**a**), 15 days (**b**), 30 days (**c**), 45 days (**d**) and 60 days (**e**).

Half a month later, most of the water areas in the bay that had not been exchanged five days ago have been exchanged to a certain extent, and the concentration value has dropped to less than 0.9 units.

One month later, except for the small tidal flats with high elevation in Xiaodao, most of the water bodies in Sanmen Bay have completed semi exchange, and the water exchange rate in the Bay has basically reached more than 60%. To facilitate the observation of the concentration distribution in the bay, the upper limit of concentration in the figure is changed to 0.4 units. The concentration isolines at the top and mouth of the bay continue to move into the bay, and the concentration gradient is significantly reduced. The overall distribution trend of the water exchange degree is still similar to that before. The concentration contour line in the western part of the bay is roughly along the NE-SW direction, and the relatively high concentration waters are mainly located in Qimen port and Haiyou harbor, with a concentration value of between 0.25 and 0.35. The concentration of Shipu port in the bay head has dropped below 0.1 unit.

Figure 8d,e shows that the trend of concentration distribution and isoline trend in the bay are close, while the overall concentration decreases and the concentration contour continues to be extrapolated. After one and a half months, the exchange ratio of all water bodies in the bay reaches more than 85%; after two months, the exchange ratio of most water bodies in the bay reaches more than 90%, except for some waters at the top of the west of the Bay. It can be considered that the water exchange in Sanmen Bay has been basically completed at this time.

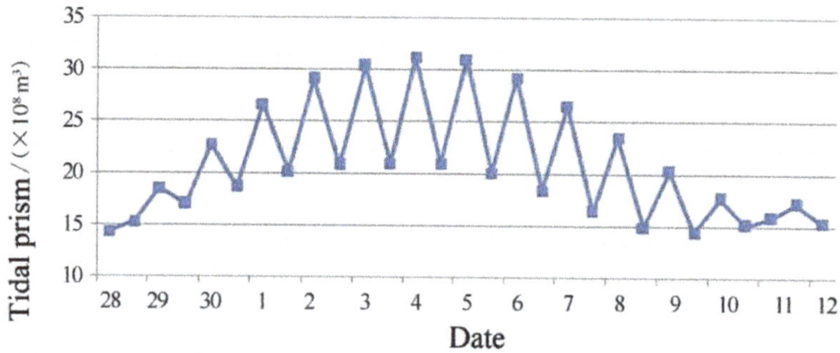

Figure 8. Tidal prism changing process (28 November 2009 to 12 December 2009).

3.3.2. Water Exchange Ability

Tidal prism means the volume of tidal water that a certain bay can hold. It is an important indicator of the environmental assessment of a bay and an important parameter reflecting the exchange of sea water in the bay. The amount of tidal prism is of great significance to the marine environment, the exchange of water bodies in the bay, the maintenance of the port branch, and the water depth in the channel. Sanmen Bay is a typical semi-enclosed bay, the rivers flowing into the bay is mountainous rivers. Therefore, when calculating the tidal volume of Sanmnen Bay, the tidal volume flowing out of the coastal estuary from the calculation area is ignored, and only the part in Sanmen Bay is considered. Two tidal current channel sections between the bay and the open sea, the Sanmen Bay section and Shipu Port are perpendicular to the longitudinal axis of the main waterway and Shipu section (Figure 8).

The tidal volume is defined as the newly added tidal volume entering the bay from the low tide time to the high tide time in any one tide cycle. The tidal volume of a tidal cycle can be expressed as:

$$Q = \int_{T_{low}}^{T_{high}} \int_{A_1} U_1 D_1 dA_1 dt + \int_{T_{low}}^{T_{high}} \int_{A_2} U_2 D_2 dA_2 dt \quad (5)$$

The hydrodynamic model is ideal for the simulation results of regional tides and tidal current processes. The simulated flow field can basically reflect the calculation of regional hydrodynamics. The calculation results can be used as the basis for the study of water exchange and water environment capacity in Sanmen Bay.

The tidal volume of Sanmen Bay is relatively large. It's between $15 \times 10^8 \sim 30 \times 10^8 m^3$ during a spring-neap tidal cycle, $20 \times 10^8 \sim 30 \times 10^8 m^3$ during spring tides, and $15 \times 10^8 \sim 18 \times 10^8 m^3$ during neap tides. The average tidal prism is about $20.78 \times 10^8 m^3$.

According to the results of numerical calculation of water exchange, the distribution of semi exchange capacity in Sanmen Bay varies greatly in different regions of the bay. Generally speaking, the water exchange capacity of the bay mouth and Shipu port is strong, and the water exchange in the west of the bay is relatively slow compared with that in the East. On the whole, the half exchange time of water in Sanmen Bay is less than 23 days, and 95% of the water exchange period is within 50 days in relatively open main water area; the half exchange time of water body in most areas of the branches is more than 10 days, and 95% of the water exchange period is more than 50 days; the water exchange time of the west end of Shipu port is longer than that of the east end, the half exchange period is less than 8 days, and 95% of the water exchange period is less than 40 days.

4. Discussion

4.1. Estimation of Environmental Capacity

Based on the precondition that to maintain the function of the environmental capacity of water, pollutant emission maximum that receiving waterbody can endure, that is to say, are permissible quantity of pollutants under the goal of water quality and hydrological condition. According to the situation of the water environment in Sanmen Bay, to calculate environment capacity about COD_{Mn} is to figure out the maximum value of emissions' sum from pollution loads at all outfalls. Gaining value which is as great as possible on environment capacity is important. Yet important management methods for contaminants controlling and environment protection might coordinate with the operability of management.

Working out a linear programming problem is to figure out a linear function's maximum or minimum under the constraint of a set of linear equations or inequations.

The response coefficient method for environment capacity calculation transforms into a linear programming maximization problem [12]. It means:

Object function:

$$max \sum_{j=1}^{n} Q_j \tag{6}$$

Constraint equation:

$$\begin{aligned} C_{0i} + \sum_{j=1}^{n} \alpha_{ij} Q_j \leq C_{si}, (i = 1, 2, \ldots, m) \\ Q_j \geq 0,\ (j = 1, 2, \ldots, n) \end{aligned} \tag{7}$$

where, j means pollutant sources' serial number, n means the number of sources; i means the serial number of control points of water quality, m means the number of control points of water quality; C_{0i} means background concentration of control points; α_{ai} means the coefficient of the jth pollution source's emission at the ith control points of water quality.

To quickly obtain the maximum value, the problem was transformed into the standard forms in linear programming. Let $C_i = C_{si} - C_{0i}$, to convert the inequation into an equivalent equation.

Object function:

$$maxQ = \sum_{j=1}^{n} Q_j \tag{8}$$

Constraint equation:

$$\sum_{j=1}^{n} \alpha_{ij} Q_j \leq C_i,\ (i = 1, 2, \ldots, m) \tag{9}$$

$$Q_j \geq 0,\ (j = 1, 2, \ldots, n) \tag{10}$$

where, C_i means the ith concentration capacity of the control points of water quality. Environment capacity assessment in this study involved COD, labile phosphate, and inorganic nitrogen. Control stations distribution can be seen in Figure 1d.

4.1.1. The Responses Factor Field of COD

According to the method of response factor, the response factor field of pollution source in each catchment unit needs to be calculated first. The response factor field uses the same region and mesh as the pollution diffusion model. The initial condition was set as 0 to exclude other source strength's influence. The concentration field (also known as response factor field) is as Table 4 shows.

Table 4. COD_{Mn} concentration field (mg/L).

Number	Bapu	Jiantiao	Haiyou	Qimen	Liyang	Baijiao	Shipu
1	0.0043	0.0035	0.0030	0.0028	0.0030	0.0019	0.0010
2	0.0011	0.0026	0.0025	0.0023	0.0025	0.0017	0.0010
3	0.0005	0.0015	0.0015	0.0015	0.0016	0.0012	0.0010
4	0.0002	0.0004	0.0005	0.0005	0.0005	0.0008	0.0012
5	0.0039	0.0046	0.0037	0.0037	0.0036	0.0021	0.0010
6	0.0010	0.0042	0.0053	0.0053	0.0054	0.0032	0.0013
7	0.0003	0.0010	0.0012	0.0012	0.0013	0.0034	0.0037
8	0.0010	0.0115	0.0097	0.0097	0.0096	0.0028	0.0012
9	0.0008	0.0040	0.0059	0.0059	0.0064	0.0053	0.0019
10	0.0007	0.0025	0.0029	0.0029	0.0029	0.0157	0.0039
11	0.0008	0.0057	0.0239	0.0239	0.0150	0.0029	0.0012
12	0.0008	0.0052	0.0108	0.0108	0.0196	0.0037	0.0014
13	0.0005	0.0020	0.0024	0.0024	0.0025	0.0281	0.0039
14	0.0005	0.0020	0.0023	0.0023	0.0024	0.0114	0.0069
15	0.0004	0.0016	0.0019	0.0019	0.0020	0.0077	0.0076
16	0.0007	0.0052	0.0273	0.0273	0.0140	0.0027	0.0011
17	0.0007	0.0051	0.0118	0.0118	0.0198	0.0028	0.0011
18	0.0004	0.0019	0.0022	0.0022	0.0023	0.0380	0.0035

4.1.2. Environmental Capacity of COD_{Mn}

To figure out COD_{Mn} permitted emission at each catchment according to Linear Programming, this study took the maximum concentration during a whole tide as background concentration to estimate the capacity. Parameters about the permitted emission of COD_{Mn} are shown in Table 5.

Table 5. Parameter on COD_{Mn} (mg/L).

Number	C_{0i}	C_{si}	Number	C_{0i}	C_{si}
1	0.952	3	10	0.827	3
2	0.944	3	11	0.852	3
3	0.921	3	12	0.844	3
4	0.882	3	13	0.815	3
5	0.892	3	14	0.820	5
6	0.857	3	15	0.830	5
7	0.852	3	16	0.877	3
8	0.842	5	17	0.849	3
9	0.837	3	18	0.811	3

Linear programming solving was conducted on 7 catchment units based on the maximum remaining capacity. The maximum remaining emission permitted can be seen in Table 6. COD_{Mn} concentration of each control station and utilization rate can be seen in Table 7. Based on the result calculated with present source strength, concentration distribution was used to estimate the emission permitted which shows that only Puba and Shipu can still discharge. The result of solving the constraint conditions of each control station is concentrated in two stations, because the calculation of linear programming is carried out according to the mathematical conditions. When selecting the largest group of total capacity among all feasible solution, it is obviously reasonable for the catchment which is relatively close to the open sea.

Table 6. Emission permitted of COD ($\times 10^4$ t/a).

Catchment	Puba	Jiantiao	Haiyou	Qimen	Liyang	Baijiao	Shipu
COD Emission Permitted	13.34	0	0	0	0	0	18.22
Overall Amount				31.56			

Table 7. Concentration of each control station concentration at the maximum environmental capacity of COD.

Control Station Number	COD Concentration	Utilization Rate	Control Station Number	COD Concentration	Utilization Rate
1	3.000	100.00	10	3.000	100.00
2	1.827	60.90	11	1.760	58.67
3	1.572	52.40	12	1.850	61.67
4	1.526	50.87	13	2.934	97.80
5	2.819	93.97	14	4.444	88.88
6	1.874	62.47	15	4.769	95.38
7	2.812	93.73	16	1.680	56.00
8	1.789	35.78	17	1.657	55.23
9	2.088	69.60	18	2.721	90.70

Table 7 shows Station 1 and Station 10 have reached the limited values. According to the prediction results of COD environmental capacity, without considering the uniformity principle, only from the aspect of maximum source strength increment, under the condition of meeting the control objectives, the maximum COD pollutant discharge capacity of Sanmen Bay is about $31.56 \times 10^4 t/a$, which is the maximum theoretical calculation result under the calculation conditions.

4.2. Reduction of Main Pollutants

As the nitrogen and phosphorus nutrients in Sanmen Bay have exceeded the standard, different emission reduction schemes for inorganic nitrogen and active phosphate should be analyzed. In this paper, the phased control method is used to calculate the pollutant reduction. Firstly, pre-calculation is conducted to analyze the influence of the variation of pollution source strength of each catchment on the distribution of the concentration field, so as to provide the basis for determining the formal calculation scheme and preliminarily determine the minimum reduction required to reach the target. Secondly, the reduction scheme is determined according to the phased control target. Finally, the results of each scheme were compared and selected, and the reduction of inorganic nitrogen pollutants in each catchment was determined on the basis of meeting the requirements of the phased control index for environmental capacity calculation of Sanmen Bay.

The water quality model was used to simulate a different reduction of inorganic nitrogen source strength, and the concentration of sea area under 0.65 mg/L was found to be relevant to the reduction of source strength [17,18]:

$$S = 336.43e^{0.9146x} \tag{11}$$

S means area, x means percentage of source strength.

According to the model result, when the recent source strength reduction reaches 14.0%, the available sea area reaches 378.16 km^2, which accounts for 60.28% of the total area in Sanmen Bay. Mid-term and long-term is long from the present, the natural conditions and socio-economic conditions may change significantly so that the prediction is likely to deviate (Figure 9). The water quality improvement target needn't be completed accurately. According to the model result, when the source strength reduction reached 30.0% and 44.0%, the available area can reach 441.72 km^2 in middle term and 497.62 km^2 in long term (Table 8).

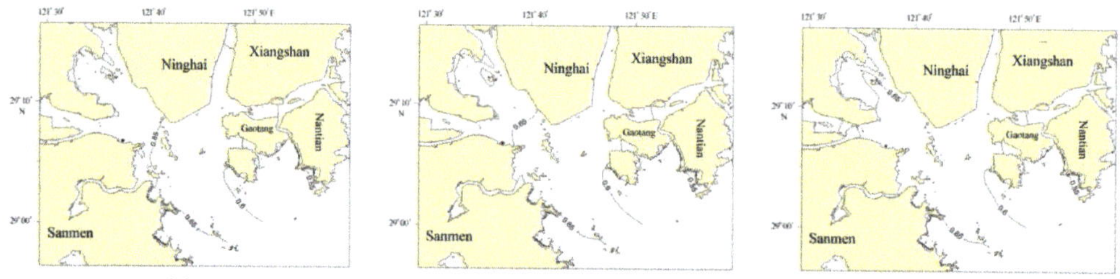

(a) (b) (c)

Figure 9. Result of source strength reduction in near term (**a**) mid-term (**b**) and long term (**c**).

Table 8. Result of source strength reduction.

Number	Reduction Rate of Source Strength/%	Area/km
1	5	356.08
2	10	369.59
3	15	384.04
4	20	399.52
5	25	421.77
6	30	445.04
7	35	461.04
8	40	480.69
9	45	510.17
10	50	536.15

5. Conclusions

A calibrated two-dimensional hydrodynamic model was built and fully validated to study the environmental characteristics of Sanmen Bay, including the tides, the residual currents, the tidal prism, and water exchange abilities.

Tides in the bay are regular semidiurnal tides, and the average tidal range is more than 4 m. The shallow water component has a certain influence on the tidal currents. The SSC in the bay is high, and is mainly caused by tidal current. The average tidal prism of the bay is about $20.78 \times 10^8 m^3$.

The distribution of semi-exchange capacity of water bodies varies greatly in different regions of the bay. Generally speaking, the water exchange capacity of the bay mouth and Shipu port is strong, and the water exchange in the west of the bay is slower than that in the East. The half exchange time of the whole bay is about 23 days, and the exchange time of 95% water body is about 60 days; the half-exchange time of relatively open sea area is less than 15 days, and 95% of water exchange time is about 50 days.

The concentrations of COD, inorganic nitrogen, and acid salt in Sanmen Bay showed a trend of being higher in the inner estuary and lower outside of the bay, and was higher in the western part and lower in the eastern part. The concentration of COD was lower than 0.60 mg/L in most areas of the eastern part of the bay, while was higher than 0.65 mg/L in the western part of the bay. The concentration of inorganic nitrogen was more than 0.70 mg/L near the west coast. The concentration of acid salt was lower in the outer bay, while was higher in the inner bay.

Author Contributions: Methodology, Y.Y. and J.Y.; software, J.Z.; validation, J.Z.; investigation, Y.Y.; resources, Y.Y.; writing—original draft preparation, J.Z.; visualization, J.Z.; supervision, L.L. and J.W. All authors have read and agreed to the published version of the manuscript.

Funding: This research was funded by the Science Technology Department of Zhejiang Province (2020C03012, 2022C03044).

Institutional Review Board Statement: Not applicable.

Informed Consent Statement: Not applicable.

Data Availability Statement: Not applicable.

Acknowledgments: This research was partially supported by a grant from the Science Technology Department of Zhejiang Province. Data were provided by Qin Chen.

Conflicts of Interest: The authors declare no conflict of interest. The funders had no role in the design of the study; in the collection, analyses, or interpretation of data; in the writing of the manuscript, or in the decision to publish the results.

References

1. Yoon, S.J.; Hong, S.; Kim, S.; Lee, J.; Kim, T.; Kim, B.; Kwon, B.-O.; Zhou, Y.; Shi, B.; Liu, P.; et al. Large-scale monitoring and ecological risk assessment of persistent toxic substances in riverine, estuarine, and costal sediments of the Yellow and Bohai seas. *Environ. Int.* **2020**, *137*, 105517.
2. Halpern, B.; Walbridge, S.; Selkoe, K.A.; Kappel, C.V.; Micheli, F.; D'agrosa, C.; Bruno, J.F.; Casey, K.S.; Ebert, C.; Fox, H.E.; et al. A global map of human impact on marine ecosystems. *Science* **2008**, *319*, 948–952.
3. Syvitski, J.; Kettner, A.; Green, P. Impact of humans on the flux of terrestrial sediment to the global coastal ocean. *Science* **2005**, *308*, 376–380.
4. Chen, B.; He, X.; Ping, Y.; Liu, L. Heavy metals pollution and potential ecological risk in sediment of Lianyungang sea area. *Mar. Environ. Sci.* **2008**, *27*, 246–249.
5. Cui, C.; Hua, W.; Yuan, G.; Jiao, X.; Fang, N.; Lv, Y.; Zhang, X.; Mao, C. Survey and evaluation on water quality of coastal area in Guanhe estuary. *China Resour. Compr. Util.* **2013**, *31*, 41–44.
6. Wu, D.; Li, W.; Xiao, R. Application of water environment capacity and total pollution quantity control to formulation of effluent criteria. *Mar. Sci. Technol.* **2005**, *28*, 48–50.
7. Linker, L.; Bautiuk, R.; Shenk, G.W.; Cerco, C.F. Development of the chesapeake bay watershed total maximum daily load allocation. *J. Am. Water Resour.* **2013**, *49*, 986–1006.
8. Liang, J.; Zhou, Y.; Wang, Z. Environmental quality assessment of the Sanmen Bay and its annual changes. *J. Zhejiang Ocean. Univ. (Nat. Sci.)* **2021**, *40*, 121–127.
9. He, Q.; Song, D.; Xu, X.-F.; Duan, Z.G. Numerical simulation of pollutants diffusion in the Sanmen Bay. *Mar. Sci. Bull.* **2018**, *37*, 63–71.
10. Sasikumar, K.; Mujumdar, P.P. Fuzzy optimization model for water quality management of a river system. *J. Water Resour. Plan. Manag.-ASCE* **1998**, *124*, 79–88.
11. Stebbing, A.R.D. Environmental Capacity and the Precautionary Principle. *Mar. Pollut. Bull.* **1992**, *24*, 287–295.
12. Subhankar, K.; Mujumdar, P.P. A two-phase grey fuzzy optimization approach for water quality management of a river system. *Adv. Water Resour.* **2007**, *30*, 1218–1235.
13. Zhang, X.; Sun, Y. Study on the environmental capacity in Jiaozhou Bay. *Mar. Environ. Sci.* **2007**, *26*, 347–359.
14. Yu, L.; Liu, S.; Wang, M. Research on SMS-Based Water Environment Capacity. *Henan Sci.* **2006**, *24*, 874–876.
15. Wang, H.; Pang, Y.; Ding, L. Calculation of the water environment capacity for a waterfront body. *Acta Sci. Circumstantiae* **2020**, *27*, 2067–2073.
16. Niu, Z.; Zhang, H. Study on estimation of coastal water environmental capacity by geostatistics and GIS. *J. Tianjin Polytech. Univ.* **2006**, *25*, 74–80.
17. Liu, H.; Yin, B. Numerical Calculation on the Content of Nitrogen, Phosphate and COD in the Liaodong Bay. *Mar. Sci. Bull.* **2006**, *25*, 46–54.
18. Su, S.; Li, H.; Xia, J. Dapeng Bay water environment capacity analysis on the base of Delft 3D model. *Res. Environ. Sci.* **2005**, *18*, 91–95.

water

MDPI

Article

Using Machine Learning Models for Predicting the Water Quality Index in the La Buong River, Vietnam

Dao Nguyen Khoi [1,2,*], Nguyen Trong Quan [3], Do Quang Linh [4], Pham Thi Thao Nhi [3] and Nguyen Thi Diem Thuy [1,2]

1 Faculty of Environment, University of Science, Ho Chi Minh City 700000, Vietnam; nguyenthidiemthuyapag@gmail.com
2 Vietnam National University, Ho Chi Minh City 700000, Vietnam
3 Institute for Computational Science and Technology, Ho Chi Minh City 700000, Vietnam; quannguyen201294@gmail.com (N.T.Q.); nhi.ptt@icst.org.vn (P.T.T.N.)
4 Institute of Hydrometeorology, Oceanology and Environment, Ho Chi Minh City 700000, Vietnam; doqlinh@gmail.com
* Correspondence: dnkhoi@hcmus.edu.vn; Tel.: +84-8-830-4379

Abstract: For effective management of water quantity and quality, it is absolutely essential to estimate the pollution level of the existing surface water. This case study aims to evaluate the performance of twelve machine learning (ML) models, including five boosting-based algorithms (adaptive boosting, gradient boosting, histogram-based gradient boosting, light gradient boosting, and extreme gradient boosting), three decision tree-based algorithms (decision tree, extra trees, and random forest), and four ANN-based algorithms (multilayer perceptron, radial basis function, deep feed-forward neural network, and convolutional neural network), in estimating the surface water quality of the La Buong River in Vietnam. Water quality data at four monitoring stations alongside the La Buong River for the period 2010–2017 were utilized to calculate the water quality index (WQI). Prediction performance of the ML models was evaluated by using two efficiency statistics (i.e., R^2 and RMSE). The results indicated that all twelve ML models have good performance in predicting the WQI but that extreme gradient boosting (XGBoost) has the best performance with the highest accuracy (R^2 = 0.989 and RMSE = 0.107). The findings strengthen the argument that ML models, especially XGBoost, may be employed for WQI prediction with a high level of accuracy, which will further improve water quality management.

Keywords: La Buong River; machine learning algorithms; surface water quality; water quality index (WQI)

Citation: Khoi, D.N.; Quan, N.T.; Linh, D.Q.; Nhi, P.T.T.; Thuy, N.T.D. Using Machine Learning Models for Predicting the Water Quality Index in the La Buong River, Vietnam. *Water* **2022**, *14*, 1552. https://doi.org/10.3390/w14101552

Academic Editor: Karl-Erich Lindenschmidt

Received: 29 March 2022
Accepted: 10 May 2022
Published: 12 May 2022

Publisher's Note: MDPI stays neutral with regard to jurisdictional claims in published maps and institutional affiliations.

1. Introduction

Surface water in rivers is a fundamental freshwater source, which plays an essential role in socio-economic development and the environment [1]. However, surface water bodies are under severe pressure because of exaggerated human activities, such as industrialization, urbanization, and population growth [2,3]. Additionally, poor management of water quantity and quality and climate change have reduced water quality during the past few decades, which leads to surface-water pollution [4,5]. Therefore, the evaluation and estimation of the water quality level in rivers are of great concern today.

The water quality index (WQI) has been extensively used to assess and classify the surface water and groundwater quality. This index by Brown et al. [6], is computed based on the physicochemical parameters of the water (e.g., temperature, pH, turbidity, dissolved oxygen (DO), biochemical oxygen demand (BOD), and concentrations of other pollutants), to estimate the level of water quality. The WQI provides quantitatively meaningful information to decision makers and planners for water resources management. However, the WQI formulations consist of lengthy calculations and thus require a lot of time and effort [5]. Additionally, the WQI formulations are inconsistent as these usually utilize different equations [7]. Accordingly, to deal with the mentioned issues, it is absolutely vital to have an alternative approach for computationally efficient and accurate estimation of the WQI.

In recent years, machine learning (ML) techniques have been extensively used for river water quality assessment, including WQI estimation [8]. These techniques have proved to be powerful tools for modeling complex non-linear behaviors in water-resource research [9]. Our literature review demonstrates that each ML algorithm has its strengths and shortcomings, and its behavior is dependent on the input variables of water quality in the different study regions. Regarding the simulation and prediction of water quality, the capability of adaptive boosting (Adaboost) [10], gradient boosting (GBM) [11], extreme gradient boosting (XGBoost) [12], decision tree (DT) [13,14], extra trees (ExT) [4], random forest (RF) [10,15], multilayer perceptron (MLP) [16], radial basis function (RBF) [17], deep feed-forward neural network (DFNN) [18], and convolutional neural network (CNN) [19] has been reported. Although there are many ML algorithms, researchers are still being confronted with problems, including which ML techniques should be applied or most appropriate for a specific problem.

In Vietnam, the WQI proposal by the Ministry of Environment and Natural Resources (MONRE) [20] requires lengthy calculations and consequently demands a lot of time and effort. However, to the best of our knowledge, no study on the use of machine learning techniques in predicting the WQI has been conducted in Vietnam. Therefore, the present study aimed to assess the performance of twelve ML algorithms, consisting of five boosting-based algorithms (Adaboost, GBM, histogram-based gradient boosting (HGBM), light gradient boosting (LightGBM), and XGBoost), three decision tree-based algorithms (DT, ExT, and RF), and four ANN-based algorithms (MLP, RBF, DFNN, and CNN), in predicting the WQI of the La Buong River in Vietnam. The La Buong River is one of the important rivers that provides water supply for domestic, agricultural, and industrial usages in the southern key economic region of Vietnam.

2. Study Area

The La Buong River (10°45′–11°00′ N, 106°50′–107°15′ E), a tributary of the Dong Nai River, has a length of approximately 56 km and a basin area of 475.8 km^2 (Figure 1). The La Buong River Basin is located in the western part of the Dong Nai province in the southern key economic region of Vietnam. The topography of the basin ranges from 10 to 385 m above sea level. The basin has a tropical monsoon climate with two different seasons: a 6-month rainy season, lasting from May to October, and a 6-month dry season, lasting from November to April. The average annual temperature was 25.4 °C, the average annual rainfall was 1786 mm, and the average annual streamflow was 7.1 m^3/s in the period 1981–2015 [21]. Rhodic Ferralsols and Ferric Acrisols are the main soils of the basin (accounting for approximately 75% of the basin area). More than 80% of land in the basin is utilized for agricultural development (cashew, coffee, and rubber). The La Buong River Basin is heavily influenced by cropping activities and livestock in the upper basin and industrial activities in the lower basin. Urbanization and industrial development are predicted to rise in the coming years [22].

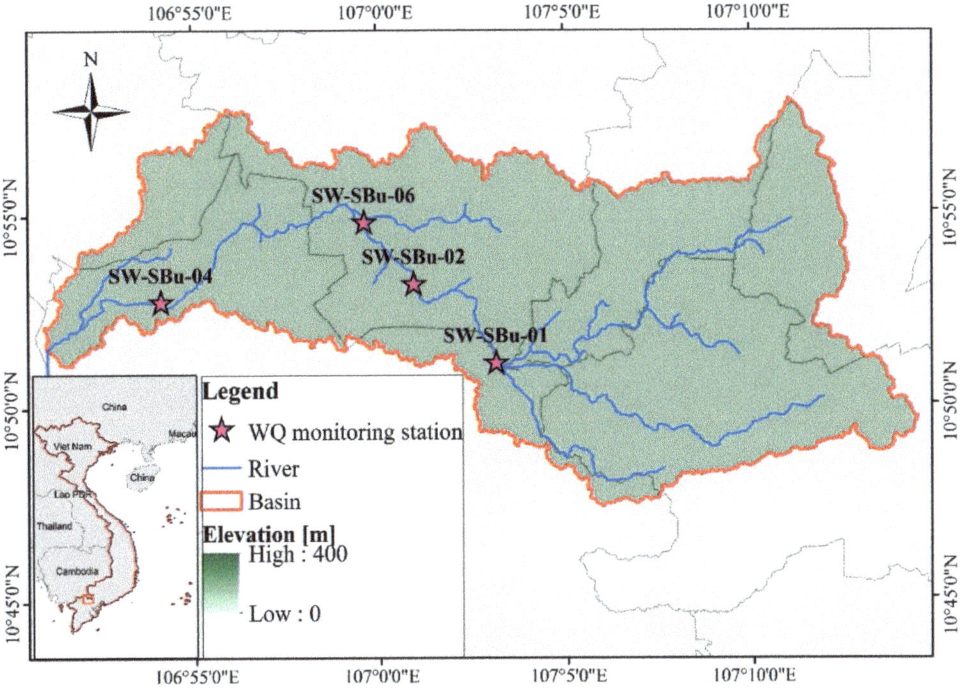

Figure 1. The La Buong River and location of the WQ monitoring stations.

3. Data and Methods

3.1. Data Collection and Processing

Eight years (2010 to 2017) of bimonthly WQ data at four WQ monitoring stations alongside the La Buong River (Figure 1) were collected from the Dong Nai Department of Natural Resources and Environment. The measured WQ data consisted of ten variables: temperature (T), pH, DO, BOD, COD, turbidity (TUR), total suspended solid (TSS), coliform, ammonium (NH_4^+), and phosphate (PO_4^{3-}). Sampling, preservation, storage, and analysis procedures followed the national guidelines for monitoring surface water.

In the current study, the ten WQ variables were utilized to compute the WQI based on Decision No. 879/QD-TCMT, issued by the Ministry of Natural Resources and Environment (MONRE) of Vietnam [20]. The WQI is expressed as follows:

$$WQI = \frac{WQI_{pH}}{100}\left[\frac{1}{5}\sum_{a=1}^{5} WQI_a \times \frac{1}{2}\sum_{b=1}^{2} WQI_b \times WQI_c\right]^{1/3} \quad (1)$$

where WQI_a is the WQI values for chemical variables (DO, BOD, COD, NH_4^+, and PO_4^{3-}), WQI_b is the WQI values for physical variables (TSS and TUR), WQI_c is the WQI value for biological variable (coliform), and WQI_{pH} is the WQI value for pH.

Based on the WQI values, the river water quality is classified into five levels: excellent (WQI = 91–100), good (WQI = 76–90), fair (WQI = 51–75), poor (WQI = 26–50), and very poor (WQI = 0–25). Full details on the guideline for calculating WQI can be found in MONRE [20]. The descriptive statistics of the WQ variables and WQI is exhibited in Table 1. The TSS, TUR, and coliform concentrations presented considerable variations, with high coefficient of variation (CV) values of 153.9% for TSS, 158.4% for TUR, and 343.2% for coliform. The high differences in these variables can be explained by the sources (point source and nonpoint source) and nature of the pollution [23]. Furthermore, the differences can be associated with seasonal effects of hydro-climatic conditions in the study area. Additionally, the WQI values indicated that the water quality of the La Buong River varies from a very low quality (WQI = 3.02) to excellent quality (WQI = 98.30).

Table 1. Descriptive statistics of the observed WQ variables and WQI in the La Buong River during 2010–2017 (n = 220).

Variables	Unit	Min	Max	Mean	Median	Std. Deviation	CV%
T	°C	25.60	32.80	28.59	28.55	1.48	5.2%
pH		5.84	8.42	7.03	7.07	0.39	5.6%
DO	mg/L	2.04	8.63	5.75	6.12	1.53	1.5%
BOD	mg/L	2.00	24.00	6.40	5.00	3.66	3.6%
COD	mg/L	3.00	113.00	19.87	16.00	14.91	15.6%
NH_4^+	mg/L	0.03	11.10	0.89	0.31	1.52	1.5%
PO_4^{3-}	mg/L	0.02	2.90	0.58	0.51	0.41	0.4%
TSS	mg/L	2.00	1402.00	85.48	31.00	156.52	153.9%
TUR	NTU	2.00	1280.00	82.36	24.00	158.36	158.4%
Coliform	MPN/100 mL	430.00	930,000.00	28,195.00	9300.00	96,766.23	343.2%
WQI		3.02	98.30	42.72	33.91	31.86	79.3%

The La Buong River WQ data were divided into two parts: 70% for the training process and 30% for the testing process. The ratio of this division is used widely in the data-driven modeling [1,7]. To improve the training speed and predictive accuracy of the ML models, the WQ data were normalized to a 0–1 range before the modeling process using the following equation:

$$x'_i = \frac{x_i - x_{min}}{x_{max} - x_{min}} \tag{2}$$

where x'_i and x_i are the normalized and original values of a WQI variable (i.e., pH, DO, BOD, etc.) at a station, and x_{min} and x_{max} are the minimum and maximum values of that variable, respectively.

3.2. Machine Learning Models

As mentioned above, the current study utilized twelve ML models for predicting WQI with three major groups: boosting-based algorithms, decision tree-based algorithms, and ANN-based algorithms.

3.2.1. Boosting-Based Algorithms

Boosting algorithm is an ensemble meta-algorithm method that aims to improve the predictive performance of several given weaker algorithms by primarily reducing bias and variance in supervised learning problems [24]. The basic principle of the boosting method starts by creating a model from the training data, and then conducting a second model based on the previous one by reducing the bias error that arises when the first model could not infer the relevant patterns in the given data. Every time a new learning algorithm is added, the weights of data are readjusted, also known as "re-weighting". These models are added sequentially until the training data is reasonably predicted or the maximum number of learners have been added to the ensemble model [25]. Five types of boosting-based algorithms were utilized in the current study, including adaptive boosting (AdaBoost), gradient boosting (GBM), histogram-based gradient boosting (HGBM), light gradient boosting (LightGBM), and extreme gradient boosting (XGBoost). Full details on these boosting-based algorithms can be found in Wu et al. [26].

3.2.2. Decision Tree-Based Algorithms

The decision tree and its many variants are the other types of learning algorithms that divide the input space into regions and has separate parameters for each region [27]. They are classified as the non-parametric supervised learning method that is widely applied for classification and regression, as well as visually and explicitly represent decisions and decision making. The typical structure of a decision tree is a tree-like flowchart, as the name goes, in which each internal node represents a "test" on an attribute, each branch represents the outcome of the test, and each leaf node represents a class label (decision taken after computing all attributes). Besides, the paths from root to leaf represent classification rules. In the present study, three decision tree-based models were assessed with respect to different learning algorithms, including decision tree (DT), extra trees (ExT), and random forest (RF). Full details on these decision tree-based algorithms can be found in Ahmad et al. [28].

3.2.3. ANN-Based Algorithms

In recent decades, AI-based models have been developed considerably to achieve a state-of-the-art architecture, comprising a number of learning algorithms and modern computational structures, across various aspects in studies on river water quality modeling [8]. ANN-based models have recently gained popularity due to its robustness and capability to handle nonlinear data even with its typically structured, single hidden layer, or advanced-structured, multiple hidden layers. Basically, ANN includes three layers: input, hidden, and output layers. In case of increasing complexity of the problem, the number of layers will rise and the computational resources will consequently also rise. In this study, both the mentioned structures of the ANN-based models were utilized for predicting WQI, such as multilayer perceptron (MLP), radial basis function (RBF), deep feed-forward neural network (DFNN), and convolutional neural network (CNN). Full details on these ANN-based algorithms can be found in Tiyasha et al. [8] and Tahmasebi et al. [29].

3.3. *Construction of ML Models*

As a first important step for constructing the ML model, the selection of input variables is required to determine a sufficient number of the variables, which have enough underlying information to predict WQI. Moreover, this selection could improve the model accuracy by avoiding the undesirable impact on the predictive performance. In the current study, ten WQ variables were identified as potential inputs. There are several existing methods to assess the input combinations, including autocorrelation function, partial autocorrelation function, cross-correlation function, and correlation coefficient. In the midst of these techniques, the correlation coefficient was selected for the current study because of its efficient and straightforward [4].

Table 2 presents that the WQ variable with the highest value of R^2 was coliform, followed by TSS, TUR, COD, BOD, PO_4^{3-}, NH_4^+, pH, DO, and T. It is noteworthy that the WQ variables of coliform, TUR, and TSS had the highest correlations with WQI due to impacts of cropping and livestock activities on water quality in the La Buong River. Based on the correlations of ten WQ variables with WQI, ten input variable combinations are listed in Table 3.

After selecting the input WQ variables, the fitted values of model parameters for each ML model were determined using a "trial and error" technique [23]. With the twelve ML models and ten scenarios of input variable combinations, 120 ML models for predicting the WQI were built during the training process and the performance of these models was evaluated during the testing process [7]. In the present study, the scikit-learn library, a Python-based package, was utilized to develop the twelve ML modes for predicting the WQI.

Table 2. Coefficient of determination (R^2) between the ten WQ variables and WQI.

Variables	T	pH	DO	BOD	COD	NH_4^+	PO_4^{3-}	TSS	TUR	Coliform
R^2	0.056	0.107	0.069	0.261	0.385	0.364	0.276	0.565	0.476	0.775

Table 3. Scenarios of input variables for the current study.

Scenarios	Input Variables
S1	Coliform
S2	Coliform, TSS
S3	Coliform, TSS, TUR
S4	Coliform, TSS, TUR, COD
S5	Coliform, TSS, TUR, COD, BOD
S6	Coliform, TSS, TUR, COD, BOD, PO_4^{3-}
S7	Coliform, TSS, TUR, COD, BOD, PO_4^{3-}, NH_4^+
S8	Coliform, TSS, TUR, COD, BOD, PO_4^{3-}, NH_4^+, pH
S9	Coliform, TSS, TUR, COD, BOD, PO_4^{3-}, NH_4^+, pH, DO
S10	Coliform, TSS, TUR, COD, BOD, PO_4^{3-}, NH_4^+, pH, DO, T

3.4. Performance Evaluation of ML Models

In the current study, two model efficiency statistics, namely, the root mean square error (RMSE) and coefficient of determination (R^2), were utilized to evaluate the goodness of fit between the predictions and observations. RMSE measures the deviation between the observed and predicted values, and R^2 measures the degree of correlation between the observed and predicted data [30].

$$\mathrm{RMSE} = \sqrt{\frac{\sum_{i=1}^{n}(O_i - P_i)^2}{n}} \quad (3)$$

$$R^2 = 1 - \frac{\sum_{i=1}^{n}(O_i - P_i)^2}{\sum_{i=1}^{n}(O_i - \overline{O})^2} \quad (4)$$

where n is the total number of predicted values, O_i is the observed value, $\overline{O}$ is the mean of observed values, and P_i is the predicted value.

4. Results and Discussion

4.1. Performance Evaluation of Boosting-Based Models

Table 4 exhibits the model performance of the boosting-based algorithms during the testing process. Results showed that AdaBoost-S2 (R^2 = 0.973 and RMSE = 0.175) had the highest performance in predicting WQI among the AdaBoost models, GBM-S7 (R^2 = 0.989 and RMSE = 0.108) had the highest performance among the GBM models, HGBM-S2 (R^2 = 0.967 and RMSE = 0.183) had the highest performance among the GBM models, LightGBM-S6 (R^2 = 0.986 and RMSE = 0.119) had the highest performance among the LightGBM models, and XGBoost-S9 (R^2 = 0.989 and RMSE = 0.107) had the highest performance among the XGBoost models under the S1–S10 scenarios. Additionally, the comparison plots of the measured WQI values with the WQI values predicted by AdaBoost-S2, GBM-S7, HGBM-S2, LightGBM-S6, and XGBoost-S9 in the testing period are shown in Figure 2. Generally, these models replicated very well the measured WQI during the testing period. However, there are small discrepancies between the measured and predicted WQI high or low values (especially those of AdaBoost-S2 and HGBM-S2). On the whole, the comparison between the boosting-based models under the S1–S10 scenarios demonstrates the XGBoost-S9 model as the best performance model.

Table 4. Efficiency statistics of the 12 ML model under the 10 scenarios of input variable combinations during the testing process.

Models		S1	S2	S3	S4	S5	S6	S7	S8	S9	S10
AdaBoost	RMSE	0.550	0.175	0.211	0.205	0.205	0.207	0.212	0.212	0.221	0.219
	R^2	0.690	0.973	0.959	0.960	0.962	0.964	0.960	0.961	0.955	0.958
GBM	RMSE	0.552	0.183	0.130	0.131	0.118	0.120	0.108	0.122	0.117	0.109
	R^2	0.682	0.967	0.983	0.983	0.986	0.986	0.989	0.986	0.987	0.989
HGBM	RMSE	0.542	0.183	0.203	0.204	0.202	0.203	0.198	0.198	0.197	0.200
	R^2	0.695	0.967	0.958	0.957	0.958	0.958	0.960	0.960	0.960	0.959
LightGBM	RMSE	0.545	0.166	0.138	0.155	0.152	0.119	0.160	0.158	0.143	0.167
	R^2	0.691	0.973	0.981	0.976	0.977	0.986	0.974	0.975	0.979	0.972
XGBoost	RMSE	0.552	0.179	0.133	0.127	0.121	0.112	0.120	0.119	0.107	0.111
	R^2	0.683	0.968	0.982	0.984	0.986	0.988	0.986	0.987	0.989	0.988
DT	RMSE	0.553	0.206	0.183	0.158	0.147	0.216	0.199	0.205	0.199	0.238
	R^2	0.681	0.957	0.966	0.976	0.979	0.954	0.960	0.957	0.960	0.941
ExT	RMSE	0.553	0.177	0.158	0.164	0.126	0.149	0.199	0.142	0.202	0.197
	R^2	0.681	0.968	0.974	0.973	0.985	0.978	0.963	0.981	0.959	0.962
RF	RMSE	0.554	0.162	0.126	0.127	0.121	0.123	0.125	0.123	0.123	0.129
	R^2	0.680	0.974	0.984	0.984	0.986	0.985	0.985	0.985	0.985	0.984
MLP	RMSE	0.532	0.153	0.192	0.132	0.141	0.196	0.928	0.307	0.996	0.515
	R^2	0.711	0.976	0.964	0.984	0.980	0.961	0.127	0.901	0.080	0.768
RBF	RMSE	0.620	0.360	0.385	0.511	0.595	0.562	0.632	0.728	0.845	0.803
	R^2	0.679	0.887	0.858	0.760	0.687	0.689	0.607	0.516	0.276	0.370
DFNN	RMSE	0.543	0.162	0.170	0.169	0.189	0.190	0.215	0.173	0.206	0.217
	R^2	0.702	0.973	0.972	0.971	0.971	0.967	0.953	0.972	0.958	0.954
CNN	RMSE	0.485	0.185	0.203	0.180	0.158	0.221	0.139	0.243	0.265	0.348
	R^2	0.773	0.965	0.962	0.964	0.977	0.961	0.982	0.942	0.937	0.895

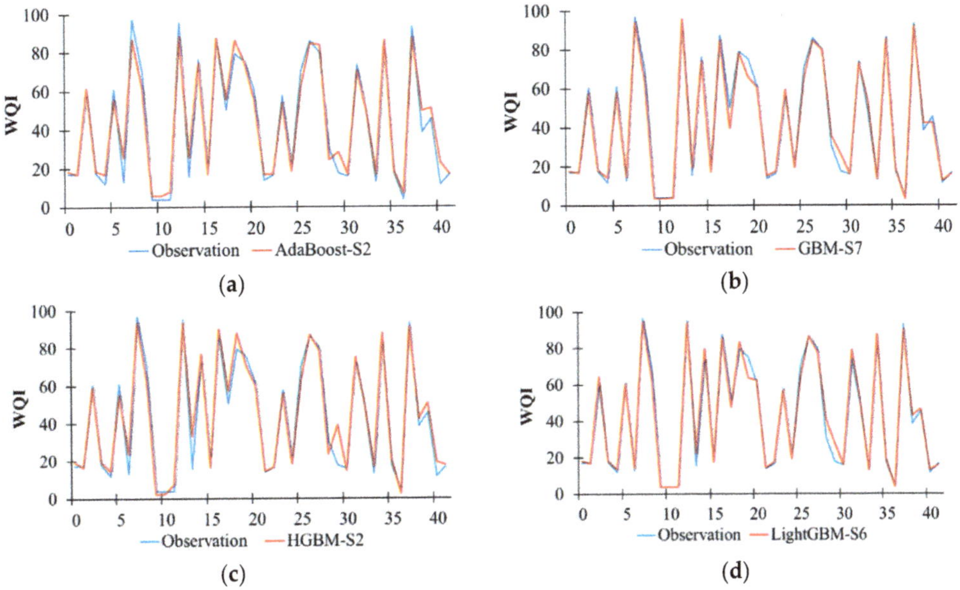

Figure 2. *Cont.*

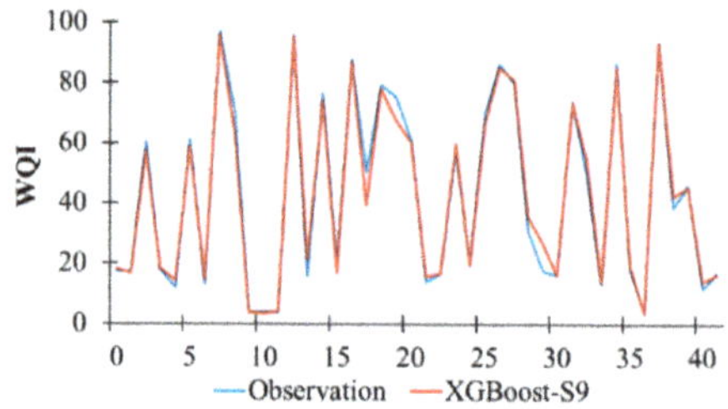

(e)

Figure 2. Temporal variation in the observed and predicted WQI values for the best performance models using boosting-based algorithms during the testing period. (**a**) AdaBoost-S2. (**b**) GBM-S7. (**c**) HGBM-S2. (**d**) LightGBM-S6. (**e**) XGBoost-S9.

4.2. Performance Evaluation of Decision Tree-Based Models

Table 4 also presents the model performance of the decision tree-based algorithms during the testing process. The results indicated that DT-S5 (R^2 = 0.979 and RMSE = 0.147), ExT-S5 (R^2 = 0.985 and RMSE = 0.126), and RF-S5 (R^2 = 0.986 and RMSE = 0.121) had the highest performance in predicting WQI among the DT models, ExT models, and RF models under the S1–S10 scenarios, respectively. Figure 3 displays the comparisons between the predicted and measured WQI for the DT-S5, ExT-S5, and RF-S5 models during the testing period. In general, all three models reproduced well the measured WQI and small differences between the measured and predicted WQI high or low values can be seen. Regarding the model performance of the decision tree-based models, RF-S5 had the highest accurate prediction.

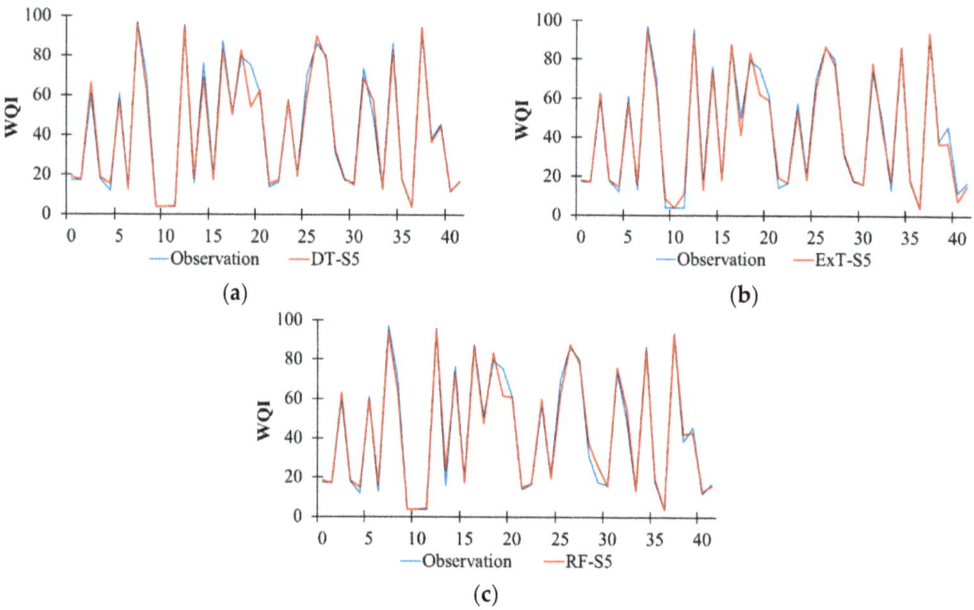

Figure 3. Temporal variation in the observed and predicted WQI values for the best performance models using decision tree-based algorithms during the testing period. (**a**) DT-S5. (**b**) ExT-S5. (**c**) RF-S5.

4.3. Performance Evaluation of ANN-Based Models

According to the model performance of the ANN-based algorithms during the testing period (Table 4), MLP-S4 (R^2 = 0.984 and RMSE = 0.132), RBF-S2 (R^2 = 0.887 and RMSE = 0.360), DFNN-S2 (R^2 = 0.973 and RMSE = 0.162), and CNN-S7 (R^2 = 0.982 and RMSE = 0.139) are the best models for predicting WQI among the MLP models, RBF models, DFNN models, and CNN models under the S1–S10 scenarios, respectively. Figure 4 illustrates the comparisons between the predicted and measured WQI for the MLP-S4, RBF-S2, DFNN-S2, and CNN-S7 models during the testing period. Generally, these four models reproduced well the measured WQI during the testing period. Moreover, small differences between the measured and predicted WQI high or low values can be observed for most models, except for RBF-S2, which show a considerable discrepancy. Regarding the model performance of the ANN-based models, MLP-S4 had the highest accurate prediction (R^2 = 0.984 and RMSE = 0.132).

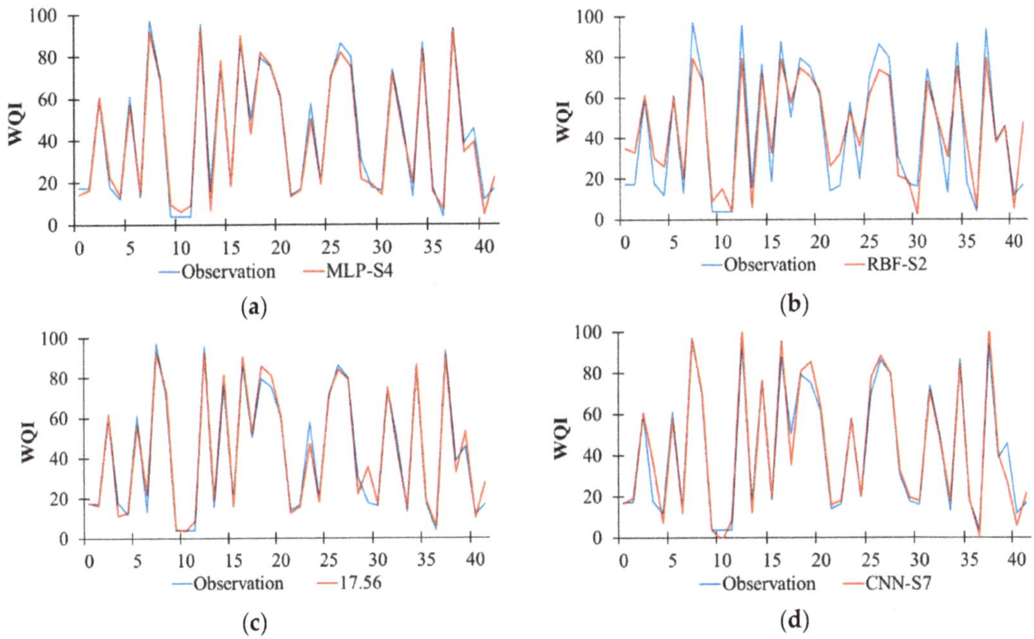

Figure 4. Temporal variation in the observed and predicted WQI values for the best performance models using ANN-based algorithms during the testing period. (**a**) MLP-S4. (**b**) RBF-S2. (**c**) DFNN-S2. (**d**) CNN-S7.

4.4. Discussion

A comparison of twelve ML models, including five boosting-based algorithms (Adaboost, GBM, HGBM, LightGBM, and XGBoost), three decision tree-based algorithms (DT, ExT, and RF), and four ANN-based algorithms (MLP, RBF, DFNN, and CNN), was conducted to evaluate their performance in predicting the WQI based on the model efficiency statistics. Based on the model performance of the twelve ML models, our findings indicate that all ML models could predict the WQI well for this study area, but the best scenarios of input variables to the ML models are different. This can be explained by the fact that each ML algorithm will respond in a different way to different input variables and data patterns [31]. As reported by Morton and Henderson [32] and Yang and Moyer [33], water quality data are characterized by a nonlinear distribution. In general, Adaboost, HGBM, RBF, and DFNN achieved good results under the S2 scenario of the input variables; DT, ExT, and RF achieved good results under the S5 scenario; and GBM and CNN achieved good results under the S7 scenario. In addition, MLP, LightGBM, and XGBoost performed well in Scenarios S4, S6, and S9, respectively. These findings indicate that most accurate prediction is dependent on the ML model parameters for the given scenario of input variables, which is consistent with results of Hussain and Khan [31].

After comparison of all twelve ML models, it indicated that the XGBoost model outperforms other ML models in the study area. In comparison with other studies, DFNN performs better than XGBoost, MLP, and RF in the Mahanadi River Basin in India [5]. Asadollah et al. [4] indicated that ExT is superior to DT and support vector regression (SVR) in the Lam Tsuen River in Hong Kong. Moreover, DT performs better as compared to the MLP model in the Rawal Dam lake in Pakistan [14]. In general, different ML algorithms will give different performance when applied to different regions. Therefore, exploring and developing a generalized ML model for applications of water quality assessment is an ongoing struggle.

As stated in previous studies, an important gap is a lack of considering cross influences between the explanatory variables, namely, the cross-correlation between land-use classes and the cross-correlation between climate conditions in influencing river water quality [34–36]. Land-use change and climate change affect hydrological components, and consequently river discharge and pollutant transport [21]. Therefore, it is essential to take into account land-use and climate changes, which may improve the accuracy of the ML models.

5. Conclusions

This research work was conducted to investigate the capability of twelve ML models, namely, five boosting-based algorithms (Adaboost, GBM, HGBM, LightGBM, XGBoost), three decision tree-based algorithms (DT, ExT, and R)), and four ANN-based algorithms (MLP, RBF, DFNN, and CNN), in predicting the WQI. The four WQ monitoring stations alongside the La Buong River were considered as a case study. Two model efficiency statistics (i.e., R^2 and RMSE) were chosen for performance comparison of the different ML models. XGBoost achieved an R^2 of 0.989 and RMSE of 0.107 in the testing process, thus being the most appropriate ML algorithm in the study area. It was followed by GBM, LightGBM, RF, ExT, MLP, CNN, DT, DFNN, AdaBoost, HGBM, and RBF. Generally, our findings strengthen the argument that ML models, particularly XGBoost, can be utilized for predicting the WQI with a high degree of accuracy, which will further improve water quality management.

Author Contributions: Conceptualization, D.N.K.; methodology, D.N.K. and N.T.Q.; software, N.T.Q. and N.T.D.T.; validation, N.T.Q. and N.T.D.T.; formal analysis, N.T.Q. and N.T.D.T.; data curation, D.Q.L. and P.T.T.N.; writing—original draft preparation, D.N.K., N.T.Q., P.T.T.N., D.Q.L. and N.T.D.T.; writing—review and editing, D.N.K., N.T.Q., P.T.T.N., D.Q.L. and N.T.D.T.; visualization, P.T.T.N. and D.Q.L.; supervision, D.N.K.; project administration, D.N.K.; funding acquisition, D.N.K. All authors have read and agreed to the published version of the manuscript.

Funding: The research was supported by the Department of Science and Technology of Ho Chi Minh City, managed by Institute for Computational Science and Technology under the contract number 11/2020/HĐ-QPTKHCN.

Institutional Review Board Statement: Not applicable.

Informed Consent Statement: Not applicable.

Data Availability Statement: Not applicable.

Acknowledgments: We would like to thank the Institute for Computational Science and Technology for supporting us to complete this research.

Conflicts of Interest: The authors declare no conflict of interest.

References

1. Nouraki, A.; Alavi, M.; Golabi, M.; Albaji, M. Prediction of water quality parameters using machine learning models: A case study of the Karun River, Iran. *Environ. Sci. Pollut. Res.* **2021**, *28*, 57060–57072. [CrossRef] [PubMed]
2. Ambade, B.; Sethi, S.S. Health Risk Assessment and Characterization of Polycyclic Aromatic Hydrocarbon from the Hydrosphere. *J. Hazard. Toxic Radioact. Waste* **2021**, *25*, 05020008. [CrossRef]
3. Ambade, B.; Sethi, S.S.; Giri, B.; Biswas, J.K.; Bauddh, K. Characterization, Behavior, and Risk Assessment of Polycyclic Aromatic Hydrocarbons (PAHs) in the Estuary Sediments. *Bull. Environ. Contam. Toxicol.* **2022**, *108*, 243–252. [CrossRef] [PubMed]
4. Asadollah, S.B.H.S.; Sharafati, A.; Motta, D.; Yaseen, Z.M. River water quality index prediction and uncertainty analysis: A comparative study of machine learning models. *J. Environ. Chem. Eng.* **2021**, *9*, 104599. [CrossRef]
5. Singha, S.; Pasupuleti, S.; Singha, S.S.; Singh, R.; Kumar, S. Prediction of groundwater quality using efficient machine learning technique. *Chemosphere* **2021**, *276*, 130265. [CrossRef]
6. Brown, R.M.; McClelland, N.I.; Deininger, R.A.; Tozer, R.G. A water quality index-do we dare. *Water Sew. Work.* **1970**, *117*, 339–343.
7. Bui, D.T.; Khosravi, K.; Tiefenbacher, J.; Nguyen, H.; Kazakis, N. Improving prediction of water quality indices using novel hybrid machine-learning algorithms. *Sci. Total Environ.* **2020**, *721*, 137612. [CrossRef]
8. Tiyasha; Tung, T.M.; Yaseen, Z.M. A survey on river water quality modelling using artificial intelligence models: 2000–2020. *J. Hydrol.* **2020**, *585*, 124670. [CrossRef]
9. Nearing, G.S.; Kratzert, F.; Sampson, A.K.; Pelissier, C.S.; Klotz, D.; Frame, J.M.; Prieto, C.; Gupta, H.V. What Role does Hydrological Science Play in the Age of Machine Learning? *Water Resour. Res.* **2021**, *57*, e2020WR028091. [CrossRef]
10. El Bilali, A.; Taleb, A.; Brouziyne, Y. Groundwater quality forecasting using machine learning algorithms for irrigation purposes. *Agric. Water Manag.* **2021**, *245*, 106625. [CrossRef]
11. Nayan, A.-A.; Kibria, M.G.; Rahman, M.O.; Saha, J. River Water Quality Analysis and Prediction Using GBM. In Proceedings of the 2020 2nd International Conference on Advanced Information and Communication Technology (ICAICT), Dhaka, Bangladesh, 28–29 November 2020; IEEE: New York, NY, USA, 2020; pp. 219–224.
12. Bedi, S.; Samal, A.; Ray, C.; Snow, D. Comparative evaluation of machine learning models for groundwater quality assessment. *Environ. Monit. Assess.* **2020**, *192*, 776. [CrossRef] [PubMed]
13. Radhakrishnan, N.; Pillai, A.S. Comparison of Water Quality Classification Models using Machine Learning. In Proceedings of the 2020 5th International Conference on Communication and Electronics Systems (ICCES), Coimbatore, India, 10–12 June 2020; IEEE: New York, NY, USA, 2020; pp. 1183–1188.
14. Ahmed, M.; Mumtaz, R.; Hassan Zaidi, S.M. Analysis of water quality indices and machine learning techniques for rating water pollution: A case study of Rawal Dam, Pakistan. *Water Supply* **2021**, *21*, 3225–3250. [CrossRef]
15. Naloufi, M.; Lucas, F.S.; Souihi, S.; Servais, P.; Janne, A.; Wanderley Matos De Abreu, T. Evaluating the Performance of Machine Learning Approaches to Predict the Microbial Quality of Surface Waters and to Optimize the Sampling Effort. *Water* **2021**, *13*, 2457. [CrossRef]
16. Gazzaz, N.M.; Yusoff, M.K.; Aris, A.Z.; Juahir, H.; Ramli, M.F. Artificial neural network modeling of the water quality index for Kinta River (Malaysia) using water quality variables as predictors. *Mar. Pollut. Bull.* **2012**, *64*, 2409–2420. [CrossRef] [PubMed]
17. Hameed, M.; Sharqi, S.S.; Yaseen, Z.M.; Afan, H.A.; Hussain, A.; Elshafie, A. Application of artificial intelligence (AI) techniques in water quality index prediction: A case study in tropical region, Malaysia. *Neural Comput. Appl.* **2017**, *28*, 893–905. [CrossRef]
18. Bowes, B.D.; Wang, C.; Ercan, M.B.; Culver, T.B.; Beling, P.A.; Goodall, J.L. Reinforcement learning-based real-time control of coastal urban stormwater systems to mitigate flooding and improve water quality. *Environ. Sci. Water Res. Technol.* **2022**. [CrossRef]
19. Prasad, D.V.V.; Venkataramana, L.Y.; Kumar, P.S.; Prasannamedha, G.; Harshana, S.; Srividya, S.J.; Harrinei, K.; Indraganti, S. Analysis and prediction of water quality using deep learning and auto deep learning techniques. *Sci. Total Environ.* **2022**, *821*, 153311. [CrossRef]
20. MONRE. *Decision No. 879/QD-TCMT on the Guidelines for Calculating Water Quality Index (WQI)*; Ministry of Natural Resources and Environment: Hanoi, Vietnam, 2011.
21. Khoi, D.N.; Nguyen, V.; Sam, T.T.; Nhi, P. Evaluation on Effects of Climate and Land-Use Changes on Streamflow and Water Quality in the La Buong River Basin, Southern Vietnam. *Sustainability* **2019**, *11*, 7221. [CrossRef]
22. Grayman, W.M.; Day, H.J.; Luken, R. Regional water quality management for the Dong Nai River Basin, Vietnam. *Water Sci. Technol.* **2003**, *48*, 17–23. [CrossRef]
23. Najah Ahmed, A.; Binti Othman, F.; Abdulmohsin Afan, H.; Khaleel Ibrahim, R.; Ming Fai, C.; Shabbir Hossain, M.; Ehteram, M.; Elshafie, A. Machine learning methods for better water quality prediction. *J. Hydrol.* **2019**, *578*, 124084. [CrossRef]
24. Zhou, Z.-H. *Ensemble Methods: Foundations and Algorithms*, 1st ed.; Chapman and Hall: Boca Raton, FL, USA, 2012.
25. Schapire, R.E. The Boosting Approach to Machine Learning: An Overview. In *Nonlinear estimation and classification*; Springer: Berlin/Heidelberg, Germany, 2003; pp. 149–171.
26. Wu, T.; Zhang, W.; Jiao, X.; Guo, W.; Hamoud, Y.A. Comparison of five Boosting-based models for estimating daily reference evapotranspiration with limited meteorological variables. *PLoS ONE* **2020**, *15*, e0235324. [CrossRef] [PubMed]

27. Geetha, A.; Nasira, G.M. Data mining for meteorological applications: Decision trees for modeling rainfall prediction. In Proceedings of the 2014 IEEE International Conference on Computational Intelligence and Computing Research, Coimbatore, India, 18–20 December 2014; IEEE: New York, NY, USA, 2014; pp. 1–4.
28. Ahmad, M.W.; Reynolds, J.; Rezgui, Y. Predictive modelling for solar thermal energy systems: A comparison of support vector regression, random forest, extra trees and regression trees. *J. Clean. Prod.* **2018**, *203*, 810–821. [CrossRef]
29. Tahmasebi, P.; Kamrava, S.; Bai, T.; Sahimi, M. Machine learning in geo- and environmental sciences: From small to large scale. *Adv. Water Resour.* **2020**, *142*, 103619. [CrossRef]
30. Krause, P.; Boyle, D.P.; Bäse, F. Comparison of different efficiency criteria for hydrological model assessment. *Adv. Geosci.* **2005**, *5*, 89–97. [CrossRef]
31. Hussain, D.; Khan, A.A. Machine learning techniques for monthly river flow forecasting of Hunza River, Pakistan. *Earth Sci. Informatics* **2020**, *13*, 939–949. [CrossRef]
32. Morton, R.; Henderson, B.L. Estimation of nonlinear trends in water quality: An improved approach using generalized additive models. *Water Resour. Res.* **2008**, *44*, W07420. [CrossRef]
33. Yang, G.; Moyer, D.L. Estimation of nonlinear water-quality trends in high-frequency monitoring data. *Sci. Total Environ.* **2020**, *715*, 136686. [CrossRef]
34. Kouadri, S.; Elbeltagi, A.; Islam, A.R.M.T.; Kateb, S. Performance of machine learning methods in predicting water quality index based on irregular data set: Application on Illizi region (Algerian southeast). *Appl. Water Sci.* **2021**, *11*, 190. [CrossRef]
35. Kung, C.-C.; Wu, T. Influence of water allocation on bioenergy production under climate change: A stochastic mathematical programming approach. *Energy* **2021**, *231*, 120955. [CrossRef]
36. Kung, C.-C.; Mu, J.E. Prospect of China's renewable energy development from pyrolysis and biochar applications under climate change. *Renew. Sustain. Energy Rev.* **2019**, *114*, 109343. [CrossRef]

water

MDPI

Article

Analysis of Seasonal Variations in Surface Water Quality over Wet and Dry Regions

Muhammad Mazhar Iqbal [1,2], Lingling Li [3,4,*], Saddam Hussain [5,6,7], Jung Lyul Lee [1,8], Faisal Mumtaz [9,10], Ahmed Elbeltagi [11], Muhammad Sohail Waqas [12] and Adil Dilawar [10,13]

1 Graduate School of Water Resources, Sungkyunkwan University 2066, Suwon 16419, Korea; mazhar0559@skku.edu (M.M.I.); jllee6359@hanmail.net (J.L.L.)
2 Water Management Training Institute (WMTI), Department of Agriculture (Water Management Wing), Government of the Punjab, Lahore 54000, Pakistan
3 Aerospace Information Research Institute, Chinese Academy of Sciences, Beijing 100101, China
4 National Engineering Laboratory for Satellite Remote Sensing Applications (NELRS), Beijing 100101, China
5 Department of Irrigation and Drainage, University of Agriculture Faisalabad, Faisalabad 38000, Pakistan; s.hussain@ufl.edu
6 Department of Biological and Agricultural Engineering, University of California, Davis, CA 95616, USA
7 Department of Agricultural and Biological Engineering, University of Florida, Gainesville, FL 32611, USA
8 School of Civil, Architecture and Environmental System Engineering, Sungkyunkwan University, Suwon 16419, Korea
9 State Key Laboratory of Remote Sensing Sciences, Aerospace Information Research Institute, Chinese Academy of Sciences, Beijing 100101, China; faisal@aircas.ac.cn
10 University of Chinese Academy of Sciences (UCAS), Beijing 100049, China; adildilawar2020@gmail.com
11 Agricultural Engineering Department, Faculty of Agriculture, Mansoura University, Mansoura 35516, Egypt; ahmedelbeltagy81@mans.edu.eg
12 Soil Conservation Wing, Punjab Agriculture Department, Murree Road, Rawalpindi 46300, Pakistan; sohailwaqasrana@gmail.com
13 State Key Laboratory of Resources and Environment Information System, Institute of Geographic Sciences and Natural Resources Research, Chinese Academy of Sciences, Beijing 100101, China
* Correspondence: lill@aircas.ac.cn

Citation: Iqbal, M.M.; Li, L.; Hussain, S.; Lee, J.L.; Mumtaz, F.; Elbeltagi, A.; Waqas, M.S.; Dilawar, A. Analysis of Seasonal Variations in Surface Water Quality over Wet and Dry Regions. *Water* **2022**, *14*, 1058. https://doi.org/10.3390/w14071058

Academic Editor: Dimitrios E. Alexakis

Received: 14 February 2022
Accepted: 19 March 2022
Published: 28 March 2022

Abstract: Water quality is highly affected by riverside vegetation in different regions. To comprehend this research, the study area was parted into wet and dry regions. The WASP8 was applied for the simulations of water quality profile over both Waterways selected from each region. It was found that the Ara Waterway, located in the wet regions, has a higher water quality variation in seasonal scale than that of the Yamuna Waterway, which is in the dry region. The interrelationship between river water quality variables and *NDVI* produce higher association for water quality variables with Pearson correlation coefficient values of about 0.66, 0.68 and −0.58, respectively, over the annual and seasonal scales in the energy limited regions. This approach will help in monitoring the seasonal variation and effect of the vegetation biomass on water quality for the sustainable water environment.

Keywords: WASP8; Ara Waterway; Yamuna River; *NDVI*; anthropogenic activities

1. Introduction

Water is one of the most important natural assets available to mankind, and its quality is entirely dependent on the local environment, utilization, treatment and reuse as per necessities and anthropogenic activities such as industrial, household, agriculture and mining operations. A rapid surge in population has exponentially intensified the demands of clean water globally [1–4]. Water quality is a key environmental issue due to its influence on aquatic life and the general health of the water ecosystem [5,6]. Generally, water quality management is performed using different water quality models of water channels and rivers, the point and nonpoint sources playing a vital role in the nitrogen dynamics, biochemical oxygen demand (BOD), and dissolved oxygen (DO) [7–10]. The surface water quality modelling can be an effective means for the simulating and predicting of contaminants fate and transport in the sustainable water environment [11,12], which can significantly save an enormous amount of labour cost, material, time, and laboratory experiments. Furthermore, it is unreachable for the in-situ sampling and experiments in some situations due to weather, hazardous, or unusual ecological issues. The simulated results of different water quality models under different contamination levels are imperative parts of different impact assessment reports and offer a base for the different environmental agencies and correct decision-making [13–15]. Hence, water quality modelling becomes a vital tool to recognize the pollution of a surface water environment, as well as the ultimate fate and behaviors of contaminants in the river environment.

Because of technological advancements, surface water quality (SWQ) modelling tools have been greatly improved in the twenty-first century, which has led to a wide range of SWQ models being developed. Several water-quality models are currently in use for decision-making and policy development in various water ecosystems around the world [16,17]. In recent years, numerous water quality models (WQM), such as the WASP model, QUASAR, and WASP-2 models, and the MIKE and BASINS models, have been developed for simulating the importance of the water quality of water bodies like rivers and lakes containing contaminants. These models were widely used in the United States, Europe, Australia, and the rest of the world [17–21]. However, WASP is the best open source model [22] for modelling the fate and transport of contaminants in rivers, lakes, and reservoirs, according to a recent analysis of the existing open source models.

The water quality factors were simulated using the SWQ model WASP (water quality simulation software). To model SWQ, the US Environmental Protection Agency (EPA) developed the WASP model [23]. The WASP models have undergone numerous iterations and have been frequently used for SWQ model evaluations during this time span [24–26]. Over the DaHan river, the WASP model has already been used [27] to simulate water quality profiles and inform policymaking. The WASP was used to model the Love River's pollutant transit and fate, and the resulting river safety plans were established [28]. For these reasons and more, a great deal of research has used the WASP model in various countries throughout the world to estimate the environmental impacts of various administrative policies and to estimate pollutant loads in order to develop long-term strategies [29,30].

The water quality has been widely affected by vegetation distribution, from naturally grown to human cultivated, which is an important determinant of river water quality and has a dynamic role to play in the water environment. Vegetation distribution can be estimated using an index called vegetation index, which is a mathematical intermingling of different spectral bands that highlights the spectral appearances of green vegetation so that they seem diverse from other image landscapes. Simple and efficient methods for descriptive and analytical assessment of green cover, robustness, and growth pattern have been developed using remote sensing-based canopies. There are various vegetation indices, with many having similar correlation and equaling functions. These indices have been broadly applied within a remote sensing application using different unmanned aerial vehicles (UAV) and satellite datasets [31,32].

The extent to which vegetation distribution hierarchically influences water environment at spatial and temporal scales is a crucial question. Dosskey et al. [33] reviewed many studies and summarized the major findings by which vegetation influences the chemistry of surface water quality, as well as how water chemistry varies among a green vegetation. All the previous studies only focus on the relationship among the water quality and vegetation indices using random water quality sampling data, which does not explain the spatial effects of vegetation on the water quality over an entire river profile. This study focuses on: (1) Analysing the seasonal variation of stream water quality over different geo-graphical regions such as energy limited (wet) and water limited (dry) regions. (2) To set up a space-time scale inter-relationship among water quality and vegetation indices.

2. Materials and Methods

2.1. Study Area

The study was conducted over energy limited and water limited regions on the basis of the global climatological aridity index (GCAI). Regions were separated using global data assimilation system (GLDAS) datasets conferring to the concept of a framework known as Budyko [34,35]. The GCAI is a numerical indicator of the degree of dryness or wetness (GCAI = PTm/Pm, where PTm = mean annual potential evapotranspiration and Pm = mean annual precipitation). The regions having an index value GCAI < 1.5 were classified as "energy limited", and those with GCAI > 1.5 as "water limited" [36]. Two Waterways were selected, one from the energy limited regions: Gyeongin Ara Waterway also known as Ara Canal, South Korea, and the other from the water limited region: Yamuna River, India. Figure 1 shows a description of the study area, Waterways, and point source locations.

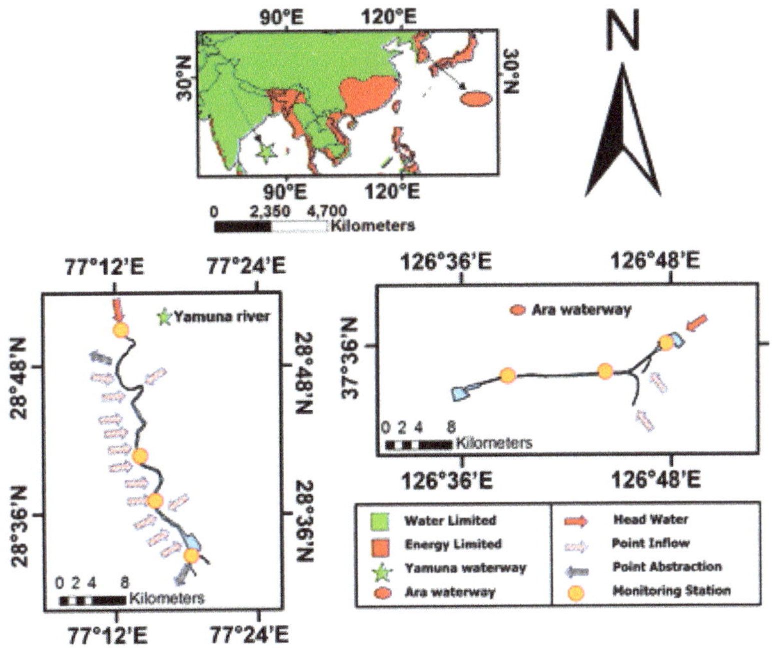

Figure 1. Study area description.

2.1.1. The Gyeongin Ara Waterway

The Gyeongin Ara Waterway is a recently constructed canal in South Korea. The main source of the flow in the Ara Waterway is the naturally sourced water from the Han River and the Gulpocheon River. The Ara Waterway was built to control flood indemnities in the basin of the Gulpocheon River and also to encourage the socio-economic growth of the area by reducing logistic expenses. The Waterway connects the Han River to the Yellow Sea. The Han River passes through Seoul and the western coastline zone of South Korea. There are two operation gates at both ends of the Waterway. The Han River side has a canal gate which is situated beside the Haengju bridge downstream of the Han River, while the Yellow Sea side has another gate that is situated at the Incheon city beside the western coastline zone of South Korea. The Ara Waterway has an overall length of about 19 km, while the width of the Waterway channel is about 85 m, and the mean depth of flowing water is 6.3 m. The head water inflow is discharged directly into the Waterway from the Han River, while there are other point sources, including water inflow from the Gulpocheon River, the irrigation dam, and the Rubber dam especially, during flood seasons.

2.1.2. The Yamuna River

The Yamuna Waterway, also primarily known as the Yamuna River, is the main source of domestic supply for New Delhi, India. It receives a huge volume of pollution in the stretch between New Delhi and Agra. From Palla, it travels about 10 km to Waziarabd, where the river's water is drained to the greatest extent possible for the city of New Delhi's domestic use. Thereafter, a small quantity of river water is observed, especially during the summer seasons.

After the Wazirabad barrage, a major drain known as the Najafghar drain enters the river. However, after the Najafghar drain, there are a total of thirteen small- to medium-sized drains that join the Yamuna River downstream. Approximately 39 km downstream of Palla, the river departs the city near the Okhla Barrage. The Yamuna basin has a total area of 9500 hectares, of which 8000 hectares are direct runoff [36]. Point sources, which discharge contaminated materials directly into the Yamuna River, are the most significant sources of Yamuna water pollution.

2.2. *Data Set*

This study used three different types of data sets, GLDAS 2.1 data, point source inflow, the initial con-centration of water quality variables, and Landsat 8 satellite data. The GLDAS (global land data assimilation system) datasets have been prepared through the collaboration of the following international agencies: NOAA, NASA, GSFC, and the NCEP. The GLDAS simulated datasets were developed to deliver medium resolution datasets by assimilating satellite-based and ground-based measurements involving the land surface models and data integration methods. Numerous land surface models have been created for the simulation of water and energy flux transferences among the ground and atmosphere interactions [37]. The water quality modelling data included point discharge, point source pollutant loadings of BOD, DO, T-N, river hydraulics characteristics data, and environmental parameters used in model simulations as the annual and seasonal bases for the year 2014. The water quality and river hydraulic characteristics data for the Yuma River, India, were obtained from various government and private agencies such as the Ministry of Environment, Upper Yamuna River Board, India River Forum, and Central Pollution Control Board. While the Water quality and hydraulic characteristics data of Ara Waterway, Republic of Korea were obtained from different Korean water and environment agencies such as the Ministry of Environment Korea, KWater, ECOREA, the national ATLAS of Korea, and Water and Environment Partnership in Asia (WEPA). The Landsat 8 satellite, cloud free images from Operational Land Imager (OLI) level-1 were utilized for the calculations of the vegetation index in all four seasons for the year 2014.

2.3. *Water Quality Modeling*

The US Environmental Protection Agency (EPA) has developed the WASP model for resolving issues of water quality. The WASP model has been continually developed over the years, allowing for greater simplicity of use and superior modelling of water quality in a wide range of water environments [24,38,39]. The WASP model is a water column and sediment-based dynamic simulation software for rivers and ponds. Toxicant transformation and advanced eutrophication are two kinetic modules in the WASP8 model employed in this investigation. One of the most complex modules, the advanced eutrophication module, incorporates many eutrophication characteristics. Many mass balance equations are included in this module to calculate the fate, transformations, phytoplankton, BOD, DO, and nitrification dynamics of pollution. This diagram shows the WASP8 model's interconnections, features, and structure (Figure 2). In order to calculate any parameters of water quality, the equation of mass balance is utilized, which is represented in Equation (1).

$$\frac{dC}{dt} = -A\frac{dUC}{dx} + \frac{d}{dx}\left(EA\frac{dC}{dx}\right) \pm SC \qquad (1)$$

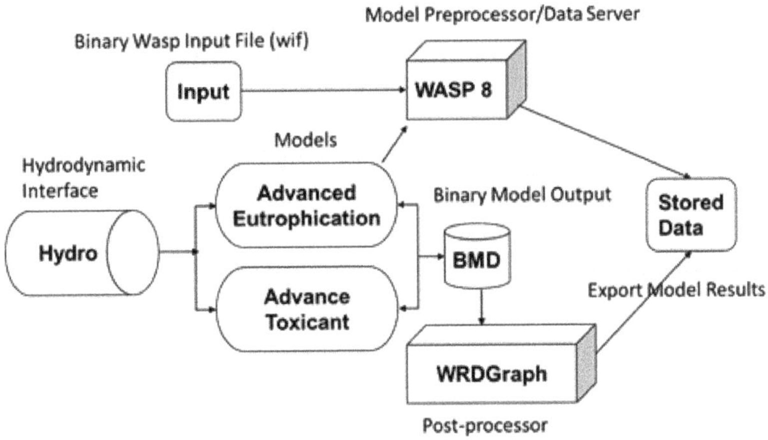

Figure 2. Components and the assembly of WASP8 model.

Whereas *C* represents the concentration of different parameters of water quality factors, *U* represents the average water velocity, *A* represents the cross-sectional area, and *x* represents the distance in one dimension. *SC* stands for exterior and internal sinks and sources, whereas *E* stands for longitudinal dispersion coefficient.

The model was developed using the required hydraulics and the water quality datasets from 2014 to 2015. The Ara Waterway longitudinal profile was equally divided into 23 segments, containing headwater inflow from the Han River and two other inflow sources from the irrigation dam and the rubber dam. The Yamuna Waterway longitudinal profile was divided into 34 segments. The inflow drains include Najafghar, Khyber Pass, Drain No 14, Magazine Road, Metcalf House, Sweeper Colony, Mori Gate, Tonga Stand, Sen Nursing, Civil Military Drain, Drain No 14, Barapulla, Power House, Maharani Bagh Drain, Hindon Cut, and the Agra Canal Abstraction. Boundaries, which included the most upstream segments of the Ara Waterway (segments 1 to 23), while the Yamuna Waterway includes point source inflow and obstructions throughout the Waterway longitudinal profile (segments 1 to 34), and was added as an initial concentration to the simulation for all the water quality variables.

2.4. Vegetation Indices

The vegetation indices are spectral transformations of two or more spectral bands combined to improve the involvement of plant properties, permit appropriate space time scale inter and intra-comparisons of global photosynthetic activities and vegetation structure distinctions. In this study, we calculated vegetation index, entitled normalized difference vegetation index (*NDVI*). A commonly applied vegetation index, the *NDVI* has long been used in ecology, remote sensing, and geography to assess the characteristics of green vegetation, including its amount (biomass), nature, and status. *NDVI* is a benchmark for spectral band ratio applications [40]. The *NDVI* monitors the vegetation state, density, and intensity of plant growth, and can be calculated from the reflectance values of the red (*RED*) and the infrared band (*NIR*). The *NDVI* values range from −1.0 to +1.0, lower values indicating sparse vegetation while higher values indicate lush green land. The *NDVI* was obtained from the reflectance values of Landsat 8 scenes by means of ArcGIS. Prepossessing of Landsat 8 scenes was performed using ENVI 5.2 software and metadata information. The following equation used to calculate the *NDVI* values was expressed as Equation (2):

$$NDVI = \frac{(NIR - RED)}{(NIR + RED)} \quad (2)$$

2.5. Model Accuracy Assesment

The model's accuracy was assessed by contrasting the actual results with those predicted by the model. It was determined that the calibrated and validated results from the prior experiments were accurate using the following five statistical estimators:

Goodness-of-fit between observed and expected data can be measured using the coefficient of determination (R^2). The R^2 has a value of 0 to 1. If R^2 is near to 1, then the model's predictions match the actual data quite well.

$$R^2 = \frac{\left\{ \sum_{i=1}^{N} (O_i - \overline{O_i})(S_i - \overline{S_i}) \right\}^2}{\sum_{i=1}^{N} (O_i - \overline{O_i})^2 \sum_{i=1}^{N} (S_i - \overline{S_i})^2} \quad (3)$$

When comparing observed and expected results, the *MAE* assesses the absolute quantitative deviation. The formula for *MAE* is as follows:

$$MAE = \frac{1}{N} \sum_{i=1}^{N} |O_i - S_i| \quad (4)$$

An index known as the mean absolute percentage error (*MAPE*) was calculated to appraise the precision of the modelled outputs. The fewer the errors, the closer the fittings of the modelled results [41], which were defined as four different stages of fitting levels, each conferring to model evaluations such as excellent, good, reasonable, and poor. If *MAPE* value <10%, fitting level is excellent, if *MAPE* value is 10–20%, fitting level is good, if *MAPE* value: 20–50%, fitting level is reasonable, and if *MAPE* value >50%, fitting level is poor.

$$MAPE = \frac{1}{N} \sum_{i=1}^{N} \left| \frac{O_i - S_i}{O_i} \right| \quad (5)$$

Here, S is a modeled outcome for a similar profile place where field observations were made for calibration and validation processes, and O is the observed value obtained from the mainstream sampling point. The total number of all the measurements were represented by N, and i is ith comparison. An observational average and a model-based simulation average are used to calculate O and S, respectively, for each location.

3. Results

3.1. Evaluation of the WASP8 Model for Its Reliability

Assessment of the WASP8 consists of a model calibration and validation investigation to establish its applicability for involvement analysis.

The WASP8 Calibrations and Validations

The surface water quality model WASP8 was primarily calibrated through the measured data attained from the four sampling locations (M1–M4) along the longitudinal stream of the Yamuna river and three sampling stations of the Ara Canal (m1–m3) Waterway on March, 2014 (spring) using the same measured coefficients (Figures 3 and 4). Then, the data collected from June (summer), September (autumn) and December (winter) were used to validate the model also by means of the same estimated reaction constants and model coefficients (Figures 5 and 6). Calibrated and validated results of the longitudinal pro-file of T-N, BOD, and DO, agreeing well with the measured values at monitoring stations and both measured and simulated, had the analogous trend of variations. The Ara Waterway shows comparatively large seasonal variation, which might be due to higher mean annual rainfall in the wetter region [42].

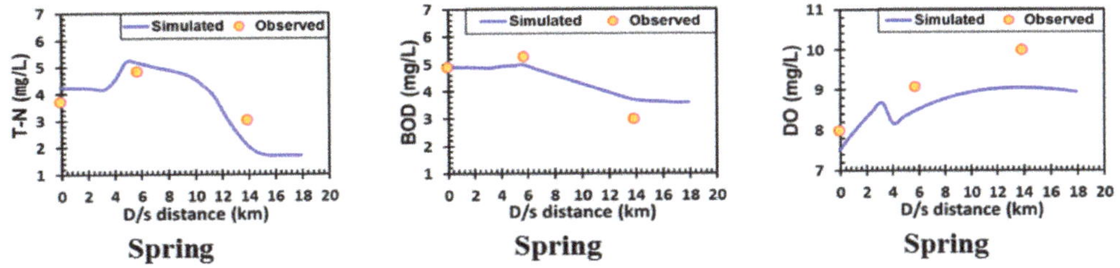

Figure 3. Calibration of water qualities in the Ara Waterway for the Spring season.

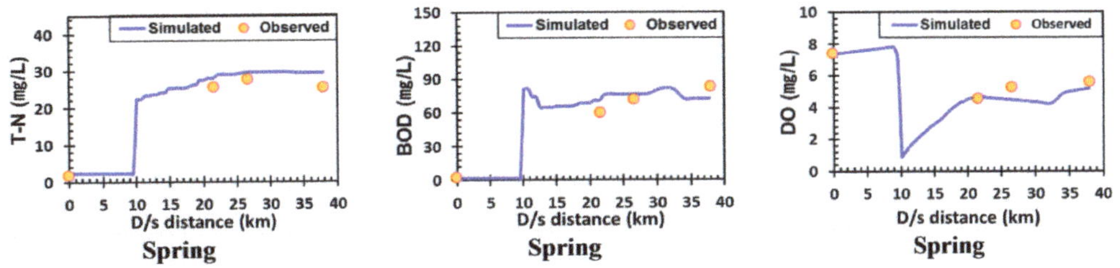

Figure 4. Calibration of water qualities in the Yamuna Waterway for the spring season.

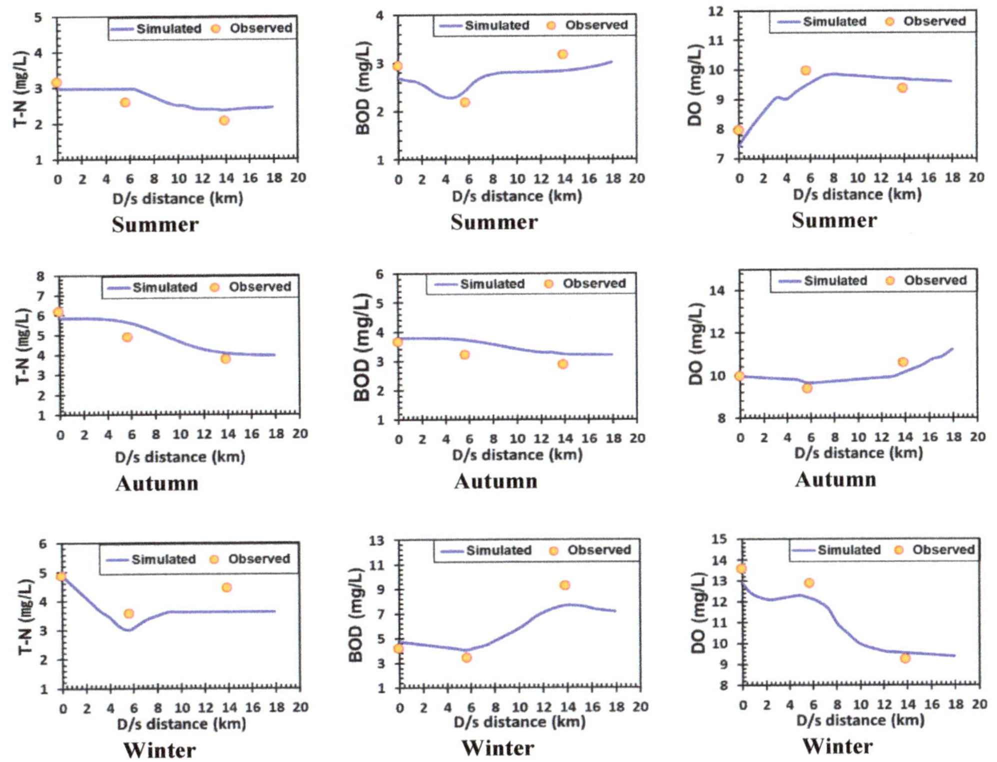

Figure 5. Confirmation of water qualities in the Ara Waterway for summer, autumn, and winter seasons.

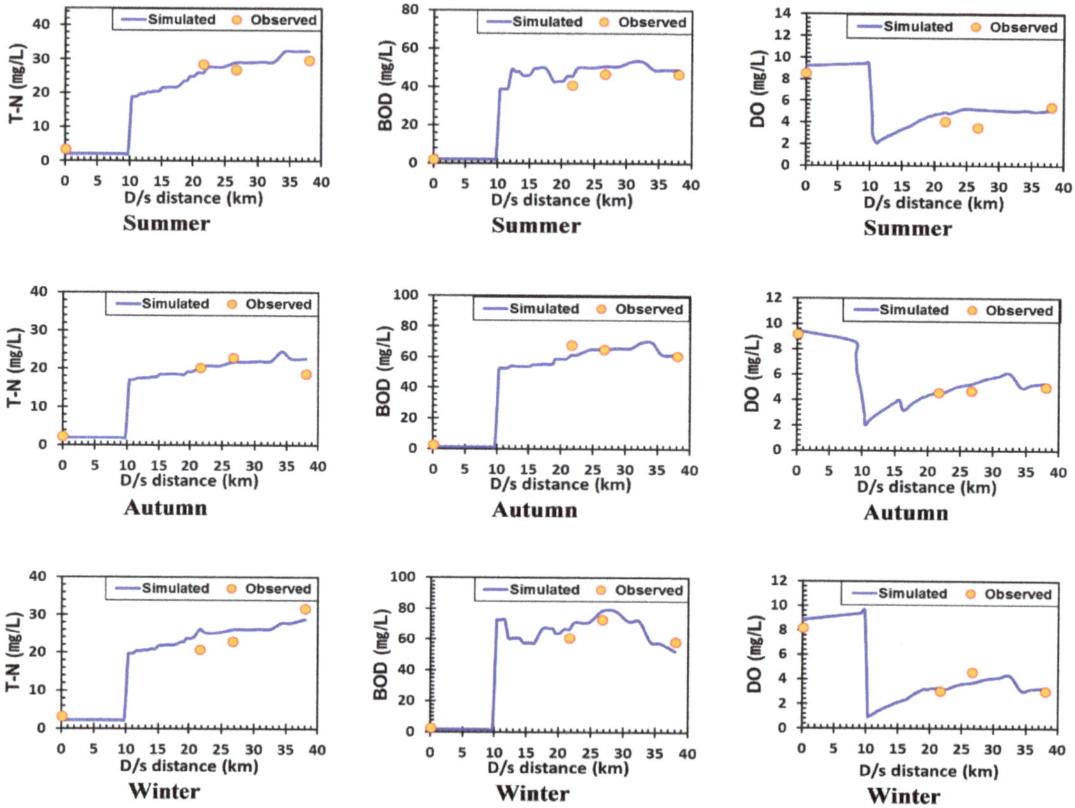

Figure 6. Confirmation of water qualities in the Yamuna Waterway for summer, autumn, and winter seasons.

The mean absolute percentage error (*MAPE*) evaluation shows that the values were enclosed among 3.5–22.61% (excellent–reasonable) (Table 1). The mean absolute error (*MAE*) shows overall estimator errors are less and agreed well with the measured data both in model calibration and validation processes (Table 2). The values of coefficient of determination R^2 had good agreement with measured data and appeared to be very close to 1 (Table 3). Moreover, all applied statistical estimators show high value of R^2 (close to 1) and lower values of errors of estimators. Overall, longitudinal profile of Yamuna Waterway shows small seasonal variation among water quality variables. At the headwater, water quality was comparatively better. However, after addition of point inflow through Najaf-garh drain water quality was further deteriorated. While, Ara Waterway showed large noticeable variation in water quality variables among each season. Total nitrogen shows a decreasing trend as it moves toward a downstream end in all seasons, while BOD and DO show further deterioration in water quality as they go toward a downstream end of the waterway profile in winter and summer seasons. In spring and autumn, BOD and DO both show improvement in water quality profile as higher rate fresh water point inflow is incorporated in waterways.

Table 1. Fitting level assessment of the WASP8 outcomes with *MAPE* process.

River	Season	T-N (%)		BOD (%)		DO (%)	
		Calibration	Validation	Calibration	Validation	Calibration	Validation
Yumna River	Spring	17.07	16.25	10.22	9.14	8.11	9.21
	Summer	10.37	9.36	8.33	8.43	6.34	6.56
	Autumn	7.54	8.34	6.91	7.25	6.08	6.73
	Winter	14.67	16.32	15.5	13.94	6.62	6.94
Ara Water Way	Spring	22.61	22.24	7.55	8.92	15.34	16.31
	Summer	9.70	9.56	5.75	4.95	8.53	8.94
	Autumn	12.4	13.1	3.50	4.31	7.41	7.64
	Winter	16.71	15.93	7.73	6.72	11.79	12.45

MAPE value, excellent; <10%, good; 10–20%, reasonable; 20–50%, bad; >50%

Table 2. Fitting level assessment of the WASP8 outcomes with *MAE* estimator.

River	Season	T-N (%)		BOD (%)		DO (%)	
		Calibration	Validation	Calibration	Validation	Calibration	Validation
Yumna River	Spring	0.15	0.16	0.62	0.58	0.12	0.13
	Summer	0.11	0.12	0.35	0.27	0.08	0.07
	Autumn	0.09	0.11	0.31	0.35	0.06	0.08
	Winter	0.17	0.15	0.78	0.66	0.14	0.12
Ara Water Way	Spring	0.13	0.14	0.44	0.51	0.10	0.11
	Summer	0.09	0.11	0.29	0.25	0.06	0.07
	Autumn	0.11	0.11	0.31	0.28	0.08	0.08
	Winter	0.16	0.14	0.52	0.47	0.13	0.11

Table 3. Fitting level assessment of the WASP8 outcomes with R^2 estimator.

River	Season	T-N (%)		BOD (%)		DO (%)	
		Calibration	Validation	Calibration	Validation	Calibration	Validation
Yumna River	Spring	0.81	0.84	0.85	0.88	0.87	0.84
	Summer	0.89	0.83	0.88	0.92	0.87	0.82
	Autumn	0.88	0.93	0.90	0.89	0.93	0.89
	Winter	0.84	0.81	0.86	0.91	0.92	0.94
Ara Water Wa	Spring	0.82	0.80	0.86	0.88	0.85	0.81
	Summer	0.87	0.94	0.90	0.89	0.91	0.88
	Autumn	0.92	0.96	0.91	0.88	0.93	0.91
	Winter	0.86	0.89	0.88	0.93	0.94	0.89

3.2. Spatial Scale Interrelationship between the Water Quality and Vegetation Indices

In the following study, we calculated vegetation index of both the Yamuna and the Ara Waterway basins at an annual and seasonal scale for the year 2014. For the case of the Ara Waterway, the vegetations are distributed sparsely along the coastline region, the Yellow Sea, while, the Yamuna Waterway has a dense vegetation beside the upstream and downstream regions. The *NDVI* values around −0.26 or less depicts a water body, values near −0.046 show snow, values near 0.002 show clouds, values near 0.025 show bare soil and values greater than 0.6 show dense vegetation. Table 4 shows the interrelationship among the vegetation index and water quality variables during the different seasons. At an annual and seasonal scale, both waterways showed higher interrelationship among vegetation index, T-N and BOD.

Table 4. The value of correlation coefficient among vegetation index and water quality variables.

Stream	Period	T−N_NDVI	BOD_NDVI	DO_NDVI
		R^2	R^2	R^2
Ara	Annual	0.66	0.68	−0.58
	Spring	0.69	0.68	−0.59
	Summer	0.52	0.62	−0.57
	Autumn	0.62	0.66	−0.42
	Winter	0.42	0.47	−0.39
Yamuna	Annual	0.55	0.51	−0.5
	Spring	0.58	0.41	−0.5
	Summer	0.42	0.48	−0.4
	Autumn	0.35	0.44	−0.39
	Winter	0.41	0.45	−0.37

However, vegetation cover has negative and the lower correlation coefficient values with DO. Overall, the water quality of the Ara Waterway has a higher correlation coefficient value than the Yamuna Waterway. From these results, we can interpret that there is some relationship between river water quality and vegetation bio-mass, and the relationship is stronger when it comes to the case of the wetter region. So, the vegetation results in more effects on stream water quality located in wetter regions.

Over basic relationships between vegetation and random point data, the use of long-term profiles provides advantages over the use of point-based water quality data. The vegetation index values take into account all green vegetation. *NDVI* values, for example, illustrate the effect of grassland, as well as agricultural and residential land values in a largely vegetated area. These values may be more or less associated with environmental conditions as a whole than they would be individually.

The *NDVI* value can, however, indicate the health of crop vegetation at various phonological growth stages, including lowered status or enhanced vegetation in a farming watershed, which may be a result of fertilizer use. Alternatively, one can examine the physical link between quality of water and a vegetation index by analyzing a regulated set of water quality and the average value of nearby vegetation indexes. Although other factors affecting the water quality include increased vegetation, greenness and moistness are related to the activities of plant physiology. Water consumption and enhanced nutrient uptake, such as oxygen and nitrogen demand, could be linked to the vegetated area, which may result in even less chemicals entering the water stream via non-point source linkage. In order to discover the root reason of the correlation between quality of water and vegetation indicators, more research is needed.

3.3. Study Limitation

In the following study, we tried to illustrate the importance of seasonal and regional based water quality management and a connection with vegetation and stream water quality. Understanding the variability in vegetation index values, of course, will depend on the mixture of different vegetation types. Because higher values of vegetation indices are not necessarily the only reason for higher interrelationship, water quality might also be affected with several degrees to different types of vegetation such as riverside forest or bushes.

Human errors introduced in this study might include errors in river boundary and general errors in the WASP modelling development. Any small variation in the segment length and vegetation index area of interest could result in a large percentage error in setting the interrelationship between the water quality and vegetation indices. In some cases, no equal number of Landsat 8 cloud free scenes is available in each season for vege-tation index acquisition. However, the study applied the average of all the available image indices for each season, and then a similar methodology was also performed with the water quality data for model simulation.

More than only vegetation biomass may have contributed to stream pollution, as evidenced by the lower R^2 values observed. Types of vegetation, land cover, soils, and other contaminant sources are all examples of these elements in play. It is possible to gain insight into the factors affecting the relations of interest by performing this type of interrelationship analysis. Future research that incorporates extensive water quality modelling with a variety of vegetation assessments and land cover classifications could be helpful in addressing this issue.

The outcomes presented in the following study showed an interrelationship among the water quality and the vegetation indices, which revealed that there must be a connection between the green vegetation biomass and the stream water quality. Furthermore, in the spring season, correlations were highest for the case of total nitrogen and biochemical oxygen demand due to the presence of higher vegetation [30]. This study also supports the theories of [31], who hypothesized that vegetation indices have the potential to some extent in illustrating stream water quality. This work is the first of its kind to use advanced water quality modelling to show how plant indices and stream water quality are linked at the spatial scale of river segments. Furthermore, this is also the first time that geographical regions were divided into a wetter and a drier region on the basis of the global climatological aridity index to investigate in a broad context.

4. Conclusions

Stream water quality was predicted using the WASP8 model in this study, which was applied to the longitudinal profile of both waterways. The modelling results were appraised through the assessment of fitting levels with *MAPE* analysis over both average and seasonal scales. The results showed a large temporal variation of the water quality concentration among seasonal scale in the Ara Waterway, which comes in the wet region, while in the dry region Yamuna Waterway showed small-scale seasonal variation in a longitudinal profile of the river water quality. Furthermore, an interrelationship was investigated among vegetation indices and water quality variables.

The Ara Waterway, which comes in wetter regions, showed relatively better connection between water quality variables and vegetation indices on the annual scale for T-N, BOD and DO with Pearson correlation coefficient values at about 0.66, 0.68 and −0.58, respectively. The seasonal scale interrelationship between water quality and *NDVI* also showed strong linkage in the energy limited region over the Ara Waterway than that of the water limited region, with a maximum correlation coefficient value occur in the spring season.

The seasonal variation of water quality variables over different geological locations will help us to understand the water quality management over different climatic regions. The management of water quality dynamics and its relationship with vegetation indices over different climatological regions will help us in monitoring and managing water quality with different strategy levels. The results of this research are advantageous for water quality management and waterway restoration, as they highlight the effects of vegetation and the relationship between water quality and vegetation indices. The variance in water quality may vary with the variation in season and vegetation biomass. Thus, better water quality of sustainable water environment could be achieved by making management strategies conferring to the different geographical region's specific vegetation patterns.

Author Contributions: Conceptualization, M.M.I., J.L.L. and S.H.; methodology, M.M.I., S.H., J.L.L., M.S.W.; software, M.M.I., S.H., J.L.L.; validation, M.M.I., S.H., J.L.L., A.D., A.E. and F.M.; formal analysis, M.M.I., M.S.W. and S.H.; investigation, M.M.I. and S.H.; resources, M.M.I., S.H., M.S.W., F.M., A.E. and L.L.; data curation, M.M.I. and S.H.; writing—original draft preparation, M.M.I. and S.H.; writing—review and editing, M.M.I., S.H., J.L.L., F.M., A.E., A.D., L.L.; visualization, M.M.I., F.M., A.D., A.E., L.L.; supervision, J.L.L., L.L.; project administration, M.M.I., S.H., M.S.W., L.L.; funding acquisition, J.L.L., S.H., L.L. All authors have read and agreed to the published version of the manuscript.

Funding: The funding is provided by the Civil Aerospace Technology Advance Research. (Grant No. Y7K00100KJ).

Institutional Review Board Statement: Not applicable.

Informed Consent Statement: Not applicable.

Data Availability Statement: Not applicable.

Acknowledgments: The authors are thankful to the Coastal and Environmental Laboratory, Graduate School of Water Resources, Sungkyunkwan University for providing resources and funding for designed research. The first and second author was supported through a scholarship funding's by the Higher Education Commission, Pakistan.

Conflicts of Interest: The authors declare no conflict of interest.

References

1. van Vliet, M.T.; Jones, E.R.; Flörke, M.; Franssen, W.H.; Hanasaki, N.; Wada, Y.; Yearsley, J.R. Global water scarcity including surface water quality and expansions of clean water technologies. *Environ. Res. Lett.* **2021**, *16*, 024020. [CrossRef]
2. Schwarzenbach, R.P.; Egli, T.; Hofstetter, T.B.; Von Gunten, U.; Wehrli, B. Global water pollution and human health. *Annu. Rev. Environ. Resour.* **2010**, *35*, 109–136. [CrossRef]
3. Choque-Quispe, D.; Froehner, S.; Palomino-Rincón, H.; Peralta-Guevara, D.E.; Barboza-Palomino, G.I.; Kari-Ferro, A.; Zamalloa-Puma, L.M.; Mojo-Quisani, A.; Barboza-Palomino, E.E.; Zamalloa-Puma, M.M. Proposal of a Water-Quality Index for High Andean Basins: Application to the Chumbao River, Andahuaylas, Peru. *Water* **2022**, *14*, 654. [CrossRef]
4. Alexakis, D.; Gotsis, D.; Giakoumakis, S. Assessment of drainage water quality in pre-and post-irrigation seasons for supplemental irrigation use. *Environ. Monit. Assess.* **2012**, *184*, 5051–5063. [CrossRef] [PubMed]
5. Arora, S.; Keshari, A.K. Dissolved oxygen modelling of the Yamuna River using different ANFIS models. *Water Sci. Technol.* **2021**, *84*, 3359–3371. [CrossRef] [PubMed]
6. Zahraeifard, V.; Deng, Z. VART model–based method for estimation of instream dissolved oxygen and reaeration coefficient. *J. Environ. Eng.* **2012**, *138*, 518–524. [CrossRef]
7. Park, S.S.; Lee, Y.S. A water quality modeling study of the Nakdong River, Korea. *Ecol. Model.* **2002**, *152*, 65–75. [CrossRef]
8. Iqbal, M.M.; Shoaib, M.; Agwanda, P.; Lee, J.L. Modeling approach for water-quality management to control pollution concentration: A case study of Ravi River, Punjab, Pakistan. *Water* **2018**, *10*, 1068. [CrossRef]
9. Khan, R.; Inam, M.A.; Iqbal, M.M.; Shoaib, M.; Park, D.R.; Lee, K.H.; Shin, S.; Khan, S.; Yeom, I.T. Removal of ZnO nanoparticles from natural waters by coagulation-flocculation process: Influence of surfactant type on aggregation, dissolution and colloidal stability. *Sustainability* **2019**, *11*, 17. [CrossRef]
10. Iqbal, M.M.; Shoaib, M.; Farid, H.U.; Lee, J.L. Assessment of water quality profile using numerical modeling approach in major climate classes of Asia. *Int. J. Environ. Res. Public Health* **2018**, *15*, 2258. [CrossRef]
11. Liu, B.; Dong, D.; Hua, X.; Dong, W.; Li, M. Spatial Distribution and Ecological Risk Assessment of Heavy Metals in Surface Sediment of Songhua River, Northeast China. *Chin. Geogr. Sci.* **2021**, *31*, 223–233. [CrossRef]
12. Huang, L.; Bai, J.; Xiao, R.; Gao, H.; Liu, P. Spatial distribution of Fe, Cu, Mn in the surface water system and their effects on wetland vegetation in the Pearl River Estuary of China. *CLEAN–Soil Air Water* **2012**, *40*, 1085–1092. [CrossRef]
13. Wang, Q.; Li, S.; Jia, P.; Qi, C.; Ding, F. A review of surface water quality models. *Sci. World J.* **2013**, *2013*, 231768. [CrossRef] [PubMed]
14. Farid, H.U.; Ahmad, I.; Anjum, M.N.; Khan, Z.M.; Iqbal, M.M.; Shakoor, A.; Mubeen, M. Assessing seasonal and long-term changes in groundwater quality due to over-abstraction using geostatistical techniques. *Environ. Earth Sci.* **2019**, *78*, 386. [CrossRef]
15. Ali, A.; Iqbal, M.; Khattak, K.K. Pilot plant investigation on the start-up of a UASB reactor using sugar mill effluent. *Cent. Asian J. Environ. Sci. Technol. Innov.* **2020**, *1*, 199–205.
16. Cox, B. A review of currently available in-stream water-quality models and their applicability for simulating dissolved oxygen in lowland rivers. *Sci. Total Environ.* **2003**, *314*, 335–377. [CrossRef]
17. Iqbal, M.M.; Shoaib, M.; Agwanda, P.O. The response of pollution load from coastal river waterfront on red tides in South Sea. *J. Coast. Res.* **2019**, *91*, 231–235. [CrossRef]
18. Fan, C.; Ko, C.-H.; Wang, W.-S. An innovative modeling approach using Qual2K and HEC-RAS integration to assess the impact of tidal effect on River Water quality simulation. *J. Environ. Manag.* **2009**, *90*, 1824–1832. [CrossRef]
19. Morley, N.J. Anthropogenic effects of reservoir construction on the parasite fauna of aquatic wildlife. *EcoHealth* **2007**, *4*, 374–383. [CrossRef]
20. Shoaib, M.; Iqbal, M.M.; Khan, R.; Lee, J.L. An Analytical Study for Eutrophication Management of Arawaterway, Korea, by Developing a Flow Model. *J. Coast. Res.* **2019**, *91*, 226–230. [CrossRef]
21. Agwanda, P.O.; Iqbal, M.M. Engineering Control of Eutrophication: Potential Impact Assessment of Wastewater Treatment Plants Around Winam Gulf of Lake Victoria in Kenya. *J. Coast. Res.* **2019**, *91*, 221–225. [CrossRef]
22. Kannel, P.R.; Kanel, S.R.; Lee, S.; Lee, Y.-S.; Gan, T.Y. A review of public domain water quality models for simulating dissolved oxygen in rivers and streams. *Environ. Modeling Assess.* **2011**, *16*, 183–204. [CrossRef]

23. Yang, C.-P.; Kuo, J.-T.; Lung, W.-S.; Lai, J.-S.; Wu, J.-T. Water quality and ecosystem modeling of tidal wetlands. *J. Environ. Eng.* **2007**, *133*, 711–721. [CrossRef]
24. Ambrose, R.B.; Wool, T.A.; Martin, J.L. *The Water Quality Analysis Simulation Program, WASP5, Part A: Model Documentation*; Environmental Research Laboratory, US Environmental Protection Agency: Athens, GA, USA, 1993.
25. Alam, A.; Badruzzaman, A.; Ali, M.A. Assessing effect of climate change on the water quality of the Sitalakhya river using WASP model. *J. Civ. Eng.* **2013**, *41*, 21–30.
26. Akomeah, E.; Chun, K.P.; Lindenschmidt, K.-E. Dynamic water quality modelling and uncertainty analysis of phytoplankton and nutrient cycles for the upper South Saskatchewan River. *Environ. Sci. Pollut. Res.* **2015**, *22*, 18239–18251. [CrossRef] [PubMed]
27. Chen, C.-Y.; Tien, C.-J.; Sun, Y.-M.; Hsieh, C.-Y.; Lee, C.-C. Influence of water quality parameters on occurrence of polybrominated diphenyl ether in sediment and sediment to biota accumulation. *Chemosphere* **2013**, *90*, 2420–2427. [CrossRef] [PubMed]
28. Yang, C.-P.; Lung, W.-S.; Kuo, J.-T.; Lai, J.-S.; Wang, Y.-M.; Hsu, C.-H. Using an integrated model to track the fate and transport of suspended solids and heavy metals in the tidal wetlands. *Int. J. Sediment Res.* **2012**, *27*, 201–212. [CrossRef]
29. Nazeer, S.; Hashmi, M.Z.; Malik, R.N. Heavy metals distribution, risk assessment and water quality characterization by water quality index of the River Soan, Pakistan. *Ecol. Indic.* **2014**, *43*, 262–270. [CrossRef]
30. Tang, P.-K.; Huang, Y.-C.; Kuo, W.-C.; Chen, S.-J. Variations of model performance between QUAL2K and WASP on a river with high ammonia and organic matters. *Desalin. Water Treat.* **2014**, *52*, 1193–1201. [CrossRef]
31. Xue, J.; Su, B. Significant remote sensing vegetation indices: A review of developments and applications. *J. Sens.* **2017**, *2017*, 1353691. [CrossRef]
32. Hussain, S.; Masud Cheema, M.J.; Arshad, M.; Ahmad, A.; Latif, M.A.; Ashraf, S.; Ahmad, S. Spray uniformity testing of unmanned aerial spraying system for precise agro-chemical applications. *Pak. J. Agric. Sci.* **2019**, *56*, 897–903.
33. Dosskey, M.G.; Vidon, P.; Gurwick, N.P.; Allan, C.J.; Duval, T.P.; Lowrance, R. The role of riparian vegetation in protecting and improving chemical water quality in streams 1. *JAWRA J. Am. Water Resour. Assoc.* **2010**, *46*, 261–277. [CrossRef]
34. Donohue, R.; Roderick, M.; McVicar, T.R. On the importance of including vegetation dynamics in Budyko's hydrological model. *Hydrol. Earth Syst. Sci.* **2007**, *11*, 983–995. [CrossRef]
35. Li, D.; Pan, M.; Cong, Z.; Zhang, L.; Wood, E. Vegetation control on water and energy balance within the Budyko framework. *Water Resour. Res.* **2013**, *49*, 969–976. [CrossRef]
36. Zhang, Y.; Leuning, R.; Chiew, F.H.; Wang, E.; Zhang, L.; Liu, C.; Sun, F.; Peel, M.C.; Shen, Y.; Jung, M. Decadal trends in evaporation from global energy and water balances. *J. Hydrometeorol.* **2012**, *13*, 379–391. [CrossRef]
37. Rodell, M.; Houser, P.; Jambor, U.; Gottschalck, J.; Mitchell, K.; Meng, C.-J.; Arsenault, K.; Cosgrove, B.; Radakovich, J.; Bosilovich, M. The global land data assimilation system. *Bull. Am. Meteorol. Soc.* **2004**, *85*, 381–394. [CrossRef]
38. Ambrose, B.; Wool, T.; Martin, J. *The Water Quality Analysis Simulation Program, WASP6, User Manua*; US EPA: Athens, GA, USA, 2001.
39. Hao, L.; Cheng, L.; Bi, X. Apply with WASP Water Quality Model. In *Proceedings of the the 2nd International Conference on Computer Application and System Modeling*; Atlantis Press: Paris, France, 2012; pp. 714–717.
40. Filippova, N.V.; Bulyonkova, T.M.; Lapshina, E. Fleshy fungi forays in the vicinities of the YSU Mukhrino field station. *Environ. Dyn. Glob. Clim. Chang.* **2015**, *6*, 3–31. [CrossRef]
41. Joyce, D.; Joyce, L.; Locke, M. *Mechanical Circulatory Support: Principles and Applications*; McGraw-Hill Education/Medical: New York, NY, USA, 2011.
42. Yeon, Y.J.; Kim, D.H.; Lee, J.L. Water quality modeling for integrated management of urban stream networks. *Int. J. Environ. Sci. Dev.* **2016**, *7*, 928. [CrossRef]

water

MDPI

Article

Application of DPSIR and Tobit Models in Assessing Freshwater Ecosystems: The Case of Lake Malombe, Malawi

Ishmael Bobby Mphangwe Kosamu [1], Rodgers Makwinja [2,3], Chikumbusko Chiziwa Kaonga [1,*], Seyoum Mengistou [2], Emmanuel Kaunda [4], Tena Alamirew [5] and Friday Njaya [6]

1 Department of Physics and Biochemical Science, Malawi University of Business and Applied Sciences, P/Bag 303, Chichiri, Blantyre 3, Malawi; ikosamu@poly.ac.mw
2 African Centre of Excellence for Water Management, Addis Ababa University, Addis Ababa P.O. Box 1176, Ethiopia; makwinjarodgers@gmail.com or makwinja.rodgers@aau.edu.et (R.M.); seyoumeng@gmail.com (S.M.)
3 Department of Fisheries, Senga Bay Fisheries Research Centre, Ministry of Forestry and Natural Resources, Salima P.O. Box 316, Malawi
4 African Centre of Excellence in Aquaculture and Fisheries (AquaFish), Lilongwe University of Agriculture and Natural Resources, Lilongwe P.O. Box 219, Malawi; ekaunda@yahoo.com or ekaunda@luanar.ac.mw
5 Water and Land Resource Centre, Addis Ababa University, Addis Ababa P.O. Box 3880, Ethiopia; tena.a@wlrc-eth.org
6 Department of Fisheries, Fisheries Deptment Headquaters, Ministry of Forestry and Natural Resources, Capitol Hill, Lilongwe P.O. Box 593, Malawi; fnjaya@gmail.com
* Correspondence: ckaonga@poly.ac.mw or ckaonga07@gmail.com; Tel.: +265-888855399

Abstract: Inland freshwater shallow lake ecosystem degradation is indistinctly intertwined with human-induced factors and climate variability. Changes in climate and human-induced factors significantly influence the state of lake ecosystems. This study provides evidence of the driver, pressure, state, impact, and response (DPSIR) indicators for freshwater lake ecosystem dynamics, taking Lake Malombe in Malawi as a case study. We used the DPSIR framework and Tobit model to achieve the study's objectives. The study's findings indicate that top-down processes gradually erode Lake Malombe's ecosystem state. The lake resilience is falling away from its natural state due to increasing rates of drivers, pressures, and impacts, indicating the lake ecosystem's deterioration. The study shows that demographic, socio-economic, climatic drivers, pressures, state, and responses significantly ($p < 0.05$) influenced the lake ecosystem's resilience. The study suggests that substantial freshwater ecosystem management under the current scenario requires a long-term, robust, and sustainable management plan. The findings from this study provide a roadmap for short-term and long-term practical policy-focused responses, particularly in implementing a freshwater ecosystem restoration programs in Malawi and Africa more broadly.

Keywords: DPSIR models; ecosystem state; Lake Malombe; tobit model

Citation: Kosamu, I.B.M.; Makwinja, R.; Kaonga, C.C.; Mengistou, S.; Kaunda, E.; Alamirew, T.; Njaya, F. Application of DPSIR and Tobit Models in Assessing Freshwater Ecosystems: The Case of Lake Malombe, Malawi. *Water* **2022**, *14*, 619. https://doi.org/10.3390/w14040619

Academic Editors: Kun Shi and Karl-Erich Lindenschmidt

Received: 14 October 2021
Accepted: 6 January 2022
Published: 17 February 2022

1. Introduction

Global inland freshwater shallow lakes are under severe stress [1,2]. Duggan et al. [3] identified ecological stressors such as pollution, climate change, over-exploitation, hydrological modification, habitat destruction, and invasion of alien species as the leading causes of freshwater ecosystem degradation. World crises such as poverty, rapid population growth, economic expansion, and climate change are also identified as the most significant emerging threats [4,5]. These identified threats are escalating such that the current scale of biodiversity loss in freshwater lakes is rapid and is considered an invisible tragedy while attracting little public, political, or scientific interest [1]. In Africa, climate change, pollution, eutrophication, and invasion of alien species are increasingly threatening the sustainability of inland freshwater shallow lakes [6–10]—impeding efforts to achieve United Nations Sustainable Development Goals (SDGs)—in particular SDG 14 (life underwater), SDG 15 (life on land), SDG 13 (climate action), SDG 1 (no poverty), SDG 6 (clean water and sanitation), SDG 2 (zero hunger), SDG 3 (good health and well-being), SDG 12 (responsible production and consumption), and Aichi Biodiversity Targets [11].

Estimates show that about 13 million hectares of global forest lands have been converted to cropland [12]. This conversion has significantly contributed to habitat degradation, eutrophication, pollution, food web alteration, and physical degradation, particularly in inland freshwater shallow lakes [13]. The Living Planet Report revealed that the global freshwater species population has declined by 83% since the 1970s, outpacing marine and terrestrial ecosystems. About 32% of the world's amphibian species are now threatened with extinction, a much higher proportion than threatened birds (12%) or mammals (23%), amouting to a total of 168 species [1,3]. About 40% of river systems globally have already been polluted, with the statistics showing an increasing trend [14]. Eutrophication, pesticides, and alteration of the biological communities continue to threaten ecosystem services' (ESs) functioning [15]. The United Nations Environmental Protection report indicates that about 30–40% of the lakes and reservoirs in the world have been affected by eutrophication [16]. Davidson & Finlayson [17] reported that 54–57% of natural wetlands had been lost, though this figure could be as high as 87%. A tremendous increase in freshwater demand, coupled with rapid population growth, has led to freshwater ecosystem crises [18]. The estimates show that about 3.6 billion people are currently facing water crises, and the projections show an increasing trend to around 5.7 billion by 2050 [19].

Although ecosystem dynamics in most African inland freshwater shallow lakes are not well documented due to a lack of data, it is very apparent that climate change and human activities have negatively affected these lakes [20]. Inland freshwater shallow lakes have been experiencing severe water level drops due to decreased precipitation and above average water temperatures [21–23]. For example, it has been reported that Lake Victoria dried up about 6000 years ago [20]. Once the most significant freshwater lake in Africa, Lake Chad has shrunk dramatically in the last 40 years due to climate variability [24]. In Kenya, Lake Naivasha, an official Ramsar Site (30,000 ha), turned into a shallow mud pool during the 2009 drought [5]. Lake Chilwa has undergone twelve recession periods between 1900 and 2012, resulting in the loss of aquatic biodiversity [25]. The ecosystem of pan-shaped Lake Chiuta in Malawi has shown a high amplification of negative consequences produced by climatic-forcing mechanisms that have impacted Lake Malombe [26]. The water levels and chemical compositions of Lake Kyoga, Lake Tana, Lake Victora, Lake Baringo, Lake Kariba, and Lake Nakuru have changed [27]. The phytoplankton production in Lake Victoria has doubled, and algal biomass has increased fourfold due to increased nutrient enrichment from human activities and industrial and domestic waste [28]. Lake Barigo's depth had decreased from 8.9 m to less than 1.9 m by 2003 due to increased human and livestock populations, coupled with deforestation [20]. Only one river is currently flowing of the seven rivers flowing into Lake Barigo in the 1970s. Alien species such as water hyacinth invaded Lake Naivasha in 1988, Lake Victoria in 1990, and is currently reported in river systems such as the Shire River that connects Lake Malawi to Lake Malombe. *Salvinia mollesta* invaded Lake Kariba in the 1970s and Lake Naivasha in the 1980s [8]. More than 50% of the endemic fish fauna in Southern Africa are listed as highly threatened categories of the IUCN [29]. There are still visible signs of water quality deterioration in many African freshwater ecosystems where rapid population growth is not matched by an expansion of ecosystem services [30].

Lake Malombe is one of the biodiversity hotspots in Malawi, accommodating various habitat types of terrestrial and aquatic flora and fauna of conservation interest [31–33]. The lake further supports various socio–economic activities such as fisheries and water supply that sustain the local population's livelihoods [34,35]. Despite its ecological and economic significance, the lake has been overstretched due to human actions, climate variability, and landscape transformation [13,26]. Due to unprecedented economic pressures, rampant sedimentation, and water quality degradation [32,36,37], increased human population and climate change have further increased the burden on the lake [38]. Attempts were made to understand the socio–ecological systems and develop a sustainable ecosystem management plan for the lake [39,40]. However, human exploitation frustrated the effort [41]. This paper aims to use the DPSIR and Tobit models to concurrently identify Lake Malombe management priorities and select appropriate monitoring indicators of the lake. The DPSIR model is a socio–economic framework that links the ecosystems' changes to the social and economic drivers and political responses [42]. It was initially developed by the Organisation for Economic Co-operation and Development (OECD) and adopted by the United Nations and the European Environmental Agency [43]. The framework detects and defines processes and interactions in human–ecological systems [44,45] and is highly linked to the ecosystem-based management approach. The DPSIR framework in this study is linked to the following research questions: "What is happening to the Lake Malombe freshwater ecosystem and why? ("D", "P", and "S"), "What are the consequences for the lake ecosystem and local population? ("I"), and "what has been done and how effective is it?" "R". However, to assess the influence of DPSIR indicators on the lake ecosystem, the following research question was framed: "To what extent do the DPSIR indicators influence the Lake Malombe ecosystem?". Since the DPSIR indicators are fragmented, the tobit model is recommended to explain the influence of these indicators on the lake's ecosystem [46].

1.1. The Theory Underpinning the DPSIR Model

The Driver–Pressure–State–Impact–Response (DPSIR) model in Figure 1 is an instrument used to detect and define processes and interactions in human–ecological systems [47].

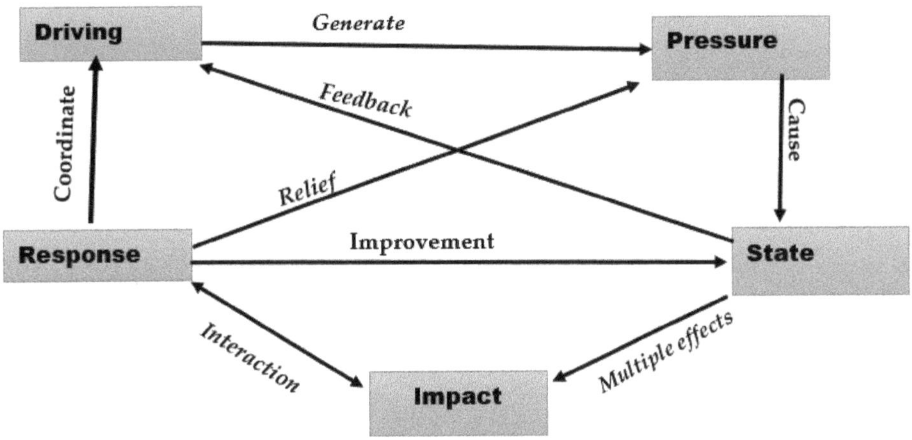

Figure 1. DPSIR conceptual framework.

It was developed in the late 1990s by the Organisation for Economic Co-operation and Development (OECD) and subsequently adopted by the United Nations and the European Environmental Agency [47]. The model provides a precise outlook in understanding the dynamics of the interface between human–natural systems and, in the process, provides a procedure for emptying and evaluating specific elements of this human–nature ecological interface in a manner that is both spatial–temporal on one hand and process-oriented on the other [48]. The DPSIR model uses similar means to other geospatial techniques focused on land–vegetation–dynamic analysis to provide an instructive podium for devaluing the intricacy of such modifications. The approach is recommended by several researchers such as Friend and Rapport [49], Borja et al. [50], Mang et al. [51], Bell [52], Pinto et al. [53], Sarmin et al. [54], Gari et al. [55], and Zare et al. [56], and is derived from PSR (pressure–state–responses)—with (D) being added later to describe underpinning drivers.

1.2. The Debate over the DPSIR as a Policy Supporting Tool

The DPSIR framework is known for its potential and usefulness in research to provide policymakers with a clear and meaningful explanation of cause–effect relationships. According to Gebremedhin et al. [57], the DPSIR model helps summarize and visualize the cause–effect interactions of the lake ecosystem in a simplified way. The Malawi government recommended the DPSIR framework in its Decentralized Environmental Management Guidelines [58]. The model extensively analyzed environmental issues originating from human activities [52]. Malawi's 2010 State of Environmental Outlook Report also adopted the integrated environmental ssessment approach based on the DPSIR model to depict environmental trend dynamics in Malawi. Other researchers have also described the model as helpful in providing the basis for policy-relevant research [59]. Although the DPSIR model has received enormous support, many heated arguments have been going on over recent decades regarding its relevance to ecosystem management, especially at the local level. Niemeijer and De Groot [60] argued that the DPSIR model focuses too much on unidirectional causalities between indicators and does not address the complex interrelationships found in reality. Other studies have noted that the DPSIR model framework does not demonstrate how different drivers and responses can be dealt with locally [61]. The model has further been disputed for failing to consider social concerns [58]. In other words, the DPSIR model ignores certain critical viewpoints and concerns of the affected groups of individuals [43]. It further displays a selective understanding of issues, making other researchers argue that the model could be biased and deemed irrelevant by some actors whose concerns are ignored [43]. Ecologists have always argued that humans protect what they value [62]. A natural ecosystem can only be protected by those who benefit from it. Individuals or groups affected by ecosystem change due to social and ecological changes have the potential to find a sustainable solution to address the impact [61]. This implies that failure to consider the concerns of many individuals could make ecosystem management complex [63]. Therefore, several researchers have suggested various approaches to improve the model to make it more flexible and easy to understand in order to communicate with various stakeholders at multiple levels. Tscherning et al. [43] suggested that incorporating social and economic concerns into the DPSIR model could make the model more effective and relevant. Svarstad et al. [63], on the other hand, suggested that various narratives need to be applied in the DPSIR model to accommodate multiple discourses of stakeholders. Niemeijer and De Groot [60] also suggested that unidirectional causal chain networks need to be replaced with multiple causal chain networks to provide a more reliable basis for selection, identify and specify indicators, and effectively help policymakers and managers to determine the precise starting points of monitoring. It is further noted that the application of the DPSIR model, along with other approaches, may provide baseline data for further research and assist researchers in framing their research objectives based on practical issues.

2. Materials and Methods

2.1. Study Area

Lake Malombe, in Figure 2 (14°40′0′′ S and 35°15′0′′ E), occupies a narrow north–south tectonic basin at the southern extension of the East African Great Rift Valley complex. It is the second-largest inland freshwater shallow lake in Malawi with a water depth of less than 6 m, length of 30 km, and width of 17 km [32]. Hydrologically, the inflow and outflow of the Shire River make the most significant contribution to Lake Malombe's water budget, except during the drought period when Lake Malawi's water level is too low to feed into Shire River [64].

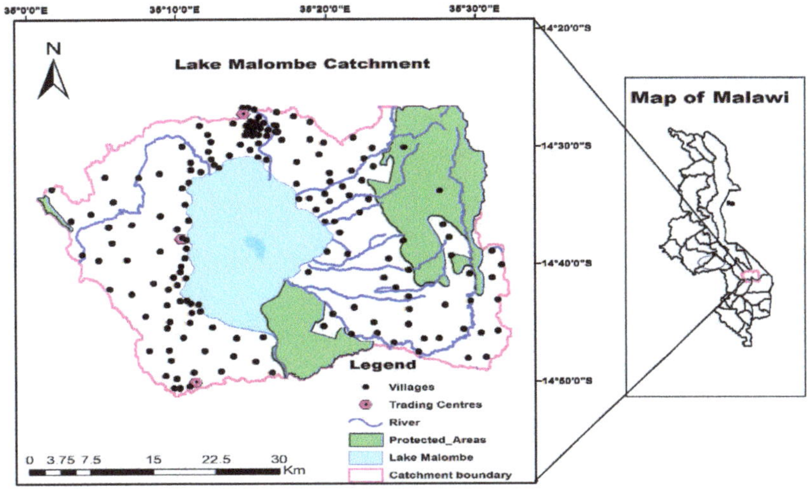

Figure 2. Map of Lake Malombe catchment (Makwinja et al., 2021a).

The rainfall period falls into three categories; early rainfall characterized by heat thunderstorms over the high ground of the escarpment area occur in October and November, the inter-tropical convergence zone (ITCZ) and the eastward encroachment of the moist Congo air mass associated with thunderstorms in December and March, and the south-easterly wind that brings rainfall over high ground in April and May. The temperature ranges from 13 °C to 35 °C, and average annual rainfall ranges from 600 mm on the rift valley floor to 1600 mm in the mountainous areas. Lake Malombe plays an essential role in the local economy and biodiversity protection. Its catchment consists of diverse habitats, including grassland, forest, shallow tidal areas, open waters, swampland, and mangroves that support diverse terrestrial and aquatic flora and fauna. The lake ecosystem supports approximately 90% of the local population in fishery activities within 10 km of the lake.

2.2. *Data Collection*

2.2.1. Grey Literature Search Strategy and Documentation

The framing of the DPSIR model indicators began with a systematic search of documentation. The DPSIR indicator selection was based on the following criteria: (1) indicator reflects ecosystem changes in state and management; (2) availability, accessibility, and consistency of indicators within the specified period of analysis; (3) independence and ability to eliminate multicollinearity; and (4) representation and superiority of the indicator over other indicators in a similar category. These criteria were achieved by searching available databases and their information within the specified period. Several scientific peer-reviewed article databases, including ScienceDirect, ResearchGate, Google Scholar, Web of Science, SciELO, LibGen, PubMed, Science-Hub, internet archives, Scopus, CrossRef, EcoPapers, BioOne, academic research databases, and others were searched. These databases were chosen based on their ability to highlight various similar peer-reviewed articles. The targeted sources were peer-reviewed journal articles, book sections, and scientific reports written in English.

We used multiple keyword combinations such as "Lake Malombe Pressure-State-Responses," "Lake Malombe Pressure-State-Impact-Responses," "Lake Malombe Driver-Pressure-State-Impact -Response," "Lake Malombe State change," "Lake Malombe Pressure-State-Response," "Lake Malombe Pressure-State-Impact Responses," "Lake Malombe Driver-Pressure-State-Effect-Action," "Lake Malombe State-Change," and "Lake Malombe Pressure-Response model". The primary database search involved screening titles and abstracts based on keywords. The open search yielded 600 scientific peer-reviewed articles, reports, and policy documents. The articles were further classified into "General" (those that were general and focused on the concept of freshwater ecosystems broadly) and "Specific" (those articles explicitly focusing on the Lake Malombe ecosystem). The articles accessed from peer-reviewed journals were further assessed based on the publisher's ranking (A* to C) where "A*" indicates highly credible sources and "C" credible sources, the relevance of the research article to the research theme, publisher identity, and reliability. These research selection criteria resulted in 105 scientific articles. The selected research articles were categorized as national (those conducted in Malawi), local (those specific to Lake Malombe), regional, and global. Data collected from the systematic review were used in qualitative and quantitative analysis.

2.2.2. Primary Data

Three traditional authorities (T/As), namely T/A Mponda, T/A Chimwala, and T/A Chowe, were ultimately selected for interviews. These three traditional authorities engage in fishing and farming and represent the two predominant Lake Malombe livelihood groups, inland and shoreline, each with distinct socio–ecological characteristics. Six university graduates coordinated face-to-face interviews held in Yao or Nyanja—native languages. The interviews were audio-recorded with participant consent and transcribed for fundamental content analysis. Semi-structured interviews were conducted with individuals in each community. Local leaders were approached first, then additional participants were engaged through the snowball sampling technique. The semi-structured format ensured that all key questions were addressed while providing ample space for the respondents to freely express their own experiences, opinions, and knowledge of the lake ecosystem DPSIR indicators. The interviews covered household activities, changes in the lake ecosystem service functions, the socio–ecological system, and associated coping mechanisms. Four focus group discussions (FGDs) per traditional authority were organized separately for households whose dominant livelihood was fishing or farming, with 10 to 12 villagers representing different socio–economic and demographic backgrounds. This design made it possible to capture a representative cross-section of residents and avoid bias through overly influential individuals. Each FGD consisted of two parts: in the first part, the participants constructed a timeline of social–ecological development from the 1970s onward, mapping and identifying critical drivers linked to ecosystem changes, pressures, current ecological conditions, livelihood activities, governance structure, and responses and ranking them based on the consensus reached. The second part involved the mental models. The mental model systematically identifies and deliberates critical events and drivers of the lake ecosystem dynamics that threaten its resilience, interaction, causes (pressures), state, consequences (impacts) for ecological and livelihood dynamics, coping mechanisms, and future policy directions. Several transect walks accompanied by villagers were also undertaken to households, local markets, fish landing sites, and farmlands to help us understand the lake ecosystem, landscape, socio–economic activities, and other livelihood practices.

The consultative meetings were conducted with the Mangochi District Environmental Office, the District Agriculture Development Office, the District Forest Office, the District Water Office, the District Disaster Risk Management Office, the District Director of Planning and Development Office, the District Fisheries Office, and the Liwonde National Park Office. Data collected from the consultative meetings helped us to develop the DPSIR framework following the Organisation for Economic Co-operation and Development (OECD) framework. The participatory rapid appraisal (PRA) approach was also used to collect data focusing on drivers, pressures, state change, impact, previous responses, and other related critical issues and was achieved through direct involvement with the local communities in the study area. The PRA encompassed a wide range of field trips, interviews, informal meetings, field observations, and in-depth key informant interviews. Before visiting the study area, a list of broad questions about the Lake Malombe ecosystem functions were framed. The duration of the field study was 30 days. The interviews and meetings were recorded and notes were taken after asking for interviewees' consent. The field trips were wrapped up by promising to send the interviewees a feedback report of the findings.

2.3. Construction of DPSIR Framework and Data Weighting

Weighting techniques were employed to select the appropriate indicators for the DPSIR framework. Specifically, entropy and analytic hierarchy techniques were used to weight indicators in different DPSIR networks. "Pressure," "state," and "response" were weighted using the hierarchy procedure. The entropy weighting technique was used to select appropriate "driving force" indicators and "impacts." The ecosystem resilience performance was demonstrated using a plot. The index of each DPSIR class was determined, after pre-determined indicator weighting, using the following specific steps [65]:

$$X'_{ij} = \frac{X_{ij} - \min(X_{1j}, X_{2j}, \ldots, X_{nj})}{\max(X_{1j}, X_{2j}, \ldots, X_{nj}) - \min(X_{1j}, X_{2j}, \ldots, X_{nj})} + 1 \quad (1)$$

$$i = 1, 2, \ldots, n; j = 1, 2 \ldots, m$$

Equation (1) demonstrates that greater the index, the more reliable it is. Conversely, Equation (2) suggests that indicators with a smaller index are best. Note that n means the number of indicators, m is a valuation indicator, and x_{ij} is an index.

$$X'_{ij} = \frac{\max(X_{1j}, X_{2j}, \ldots, X_{nj}) - X_{ij}}{\max(X_{1j}, X_{2j}, \ldots, X_{nj}) - \min(X_{1j}, X_{2j}, \ldots, X_{nj})} + 1 \quad (2)$$

$$i = 1, 2, \ldots, n; j = 1, 2 \ldots, m$$

The next step is to calculate the proportion of the i-th plan under the j-th indicator:

$$\rho_{ij} = \frac{\left(X'_{ij}\right)}{\sum_{i=1}^{n} X'_{ij}} + 1, \quad j = 1, 2, \ldots, m \quad (3)$$

The entropy of the j-th indicator is calculated as:

$$e_i = \frac{1}{\ln n_{i-1}} \sum_{i=1}^{n} p_{ij} \ln p_{ij} \quad (4)$$

The weight indicator is calculated as:

$$W_i = \frac{1 - e_i}{\sum_{j=1}^{n} (1 - e_i)}, \; j = 1, 2, \ldots, m \quad (5)$$

The comprehensive score of each indicator is determined as:

$$S_i = \sum_{j=1}^{m} W_j * \rho_{ij}, \quad 1, 2, \ldots, n \quad (6)$$

2.4. Tobit Model

James Tobit introduced the Tobit model in 1958. The model assumes that many variables have a lower or upper limit and takes these as limiting values. For example, the decision as to whether the Lake Malombe ecosystem is changing may be characterized as a dichotomous choice between two mutually exclusive alternatives (negative or positive). A breakpoint in the explanatory variables could depict this. In the Tobit model, Y denotes that the change in the lake ecosystem function depends on DPSIR indicators and could be expressed as x vectors meaning explanatory variables. The Y variable is defined in the context of a limited variable that takes on two values; $Y \sim Y^*$ if the results of the ecosystem change are positive and Y $\sim$ 0 if the results are negative. It should be noted that the probability y takes only values of 1 or 0 [65] and the thresholds of all members in the sample are distributed among different values of the explanatory variables. This can be demonstrated by a sigmoid curve (Figure 3).

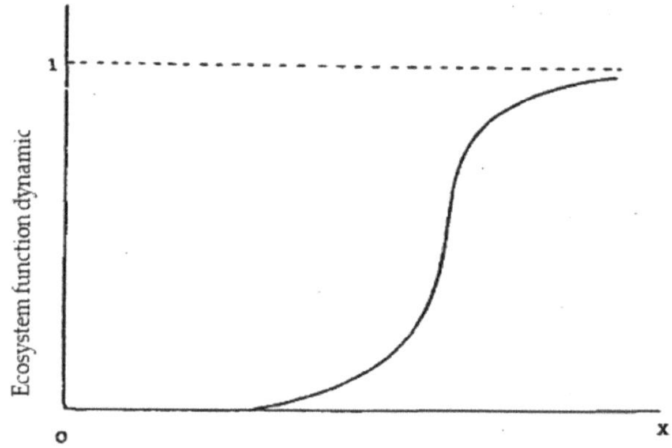

Figure 3. A sigmoid curve of Lake Malombe ecosystem function dynamics.

The explanatory variables (DPSIR indicators) considered in the Tobit model include a demographic driver (Dd), socio–economic drivers (Ds), climatic drivers (Dc), pressures (P), socio–economic states (Ss), environmental states (Se), impacts (I), capacity building and governance responses (RCBG), institutional measure responses (RIM), livelihood intervention responses (RLI), and environmental intervention responses (REI). Having clarified these concepts, we define a Tobit model [66] as follows:

$$Y^* = X'\beta + \varepsilon,\ with\ \varepsilon \,\Big|\, X \sim N\left(0;\delta^2\right), with\ y = y^* if\ y^* > 0,\ and\ y = 0\ otherwise$$

where y is the observed variable of interest, and Y^* is a latent variable. The above equation describes three things. First, the expected effect of x on Y^* is monotonic. Second, the residuals follow a normal distribution. Lastly, the dependent variable is left censored.

2.5. Data Analysis

Data were analyzed using a critical discourse approach and content analysis. The DPSIR framework guided the identification, structuring, and representation of system elements and their interactions. Field recordings were decoded and notes were scanned to identify, classify and group various stakeholders' ideas and concepts. Metadata from primary sources were exported in STATA version 15 and analyzed using descriptive and inferential statistics.

3. Results

This study analyzed the Lake Malombe freshwater ecosystem using the DPSIR model (Figure 4). The primary goal was to provide concise and specific information about the driving forces, socio–economic and environmental pressures, socio–economic and environmental states, impacts resulting from spatial and temporal changes of the lake, and responses.

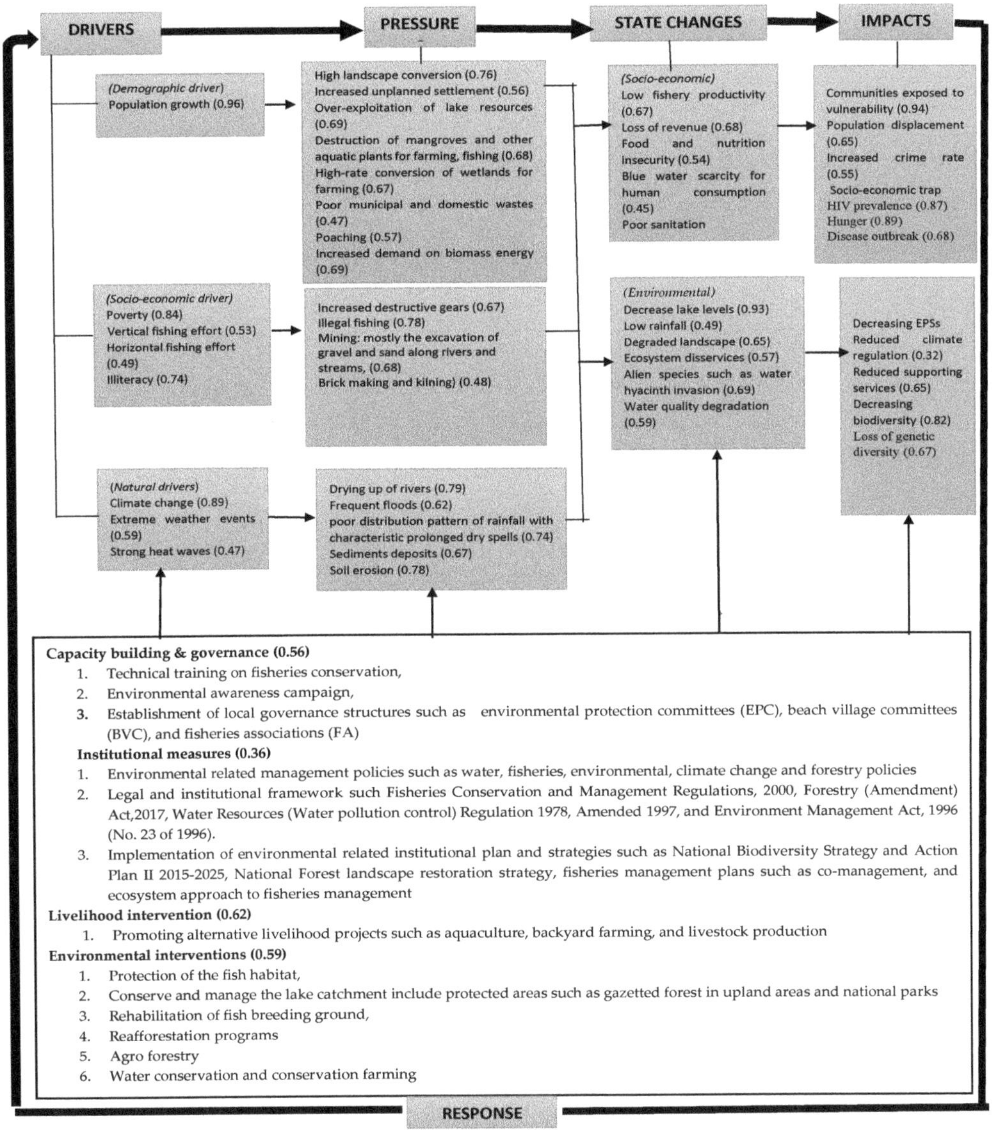

Note: The values represent weight valuation index (rated at the scale of 0 to 1 where 0 means least essential and 1 the most significant).

Figure 4. DPSIR model for Lake Malombe Inland freshwater shallow lake.

3.1. Drivers and Pressures

The driving force is the root cause for the spatial and temporal changes of the lake ecosystem state (Figure 4). It can be interpreted as a demographic, socio–economic, or natural factor.

The stakeholders ranked rapid population growth as the highest indicator for the demographic driver (0.96). The pressure associated with this driver includes high landscape conversion, increased unplanned settlement, over-exploitation of lake resources, destruction of mangroves and other aquatic plants for farming, fishing, high-rate conversion of wetlands into farmlands, buildup of municipal and domestic waste, poaching, and increased demand for biomass energy (Figure 5).

Note: Photo (a) shows a man carrying charcoal to sell at Mangochi town; photo (b) indicates the deforested area; photo (c) shows the sand extracted from the rivers ready for sale; and photo (d) indicates the burning of forest land for farming and charcoal production

Figure 5. Pressure on the Lake Malombe ecosystem.

Poverty was the highest (0.84) socio–economic driver, followed by illiteracy (0.74). Vertical fishing efforts (0.53) and horizontal fishing efforts (0.49) ranked the lowest. The pressures triggered by these socio–economic drivers include illegal fishing (0.78) (ranked the highest), gravel and sand mining along rivers and streams (0.68) (Figure 5), increased destructive gears (0.67), brick making, and kilning (0.48). Climate change was ranked the highest (0.89) natural driver, with extreme weather events (0.59) and strong heat waves (0.47) ranked the lowest. The pressures triggered by these drivers include the drying up of rivers (0.79), frequent floods (0.62), a poor rainfall distribution pattern characterized by prolonged dry spells (0.74), soil erosion (0.78), and sediment deposits (0.67), as expressed by the following anecdote: "During the intense precipitation, debris from the upland areas are washed away into the lake damaging the mangroves and contribute to the lake siltation."an older man at Chimwala Village, Nov 2019. Figure 6 shows comprehensive indices for drivers and pressures exerted on the Lake Malombe ecosystem from 1970 to 2020.

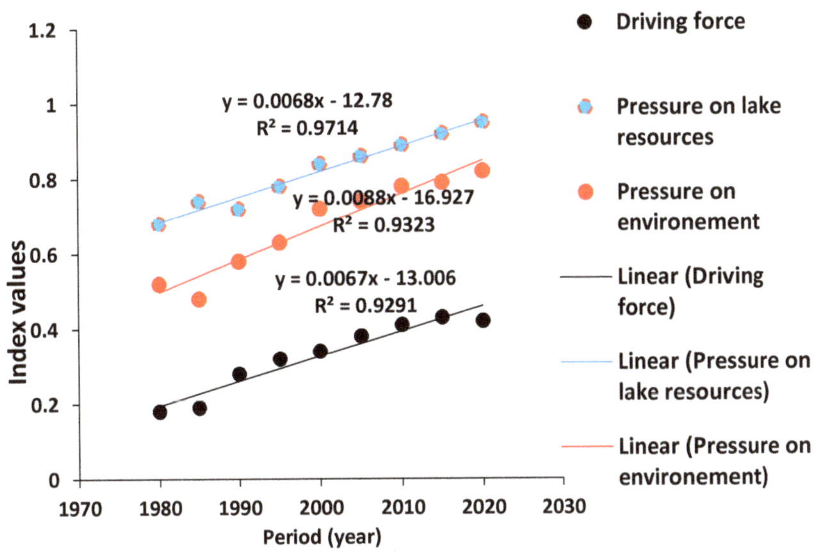

Figure 6. Inter-annual variation of the driver and pressure index.

The inter-annual variation index followed the linear pattern demonstrated by the linear regression coefficients (R squared = 0.9). The theoretical explanation of the current observation is that these drivers and pressures will continue influencing the spatial and temporal changes of the ecosystem.

3.2. State and Impacts

The state index refers to the condition of the lake ecosystem resulting from drivers such as demographics, socio–economic factors, and natural factors and pressures such as environmental factors. The socio–economic states comprised those factors affecting the local population's well-being. The stakeholders mentioned low fishery productivity (0.67), disease outbreak (0.87), loss of revenue (0.68), and food and nutrition insecurity (0.54) as the socio–economic states. The environmental states comprised factors such as decreased lake levels (0.93), water quality degradation (0.59), low rainfall (0.49), degraded landscape (0.65), ecosystem disservices (0.57), and alien species such as water hyacinth invasion (0.69) that affect the lake ecosystem function. Lake ecosystem resilience changes impact the local population and the environment. In this study, the impact index is defined as the socio–economic and environmental impacts that affect the local population's livelihood, social welfare, and the environment due to the changing state of the lake ecosystem. The impacts in this study were linked to socio–economic factors, which include communities' exposure to vulnerability (0.94), population displacement (0.65), increased crime rate (0.55), socio–economic traps, HIV prevalence (0.87), hunger (0.89), and disease outbreak (0.68). The environmental state is linked to impacts such as decreasing ecosystem provisioning services (EPSs), reduced climate regulation (0.32), reduced supporting services (0.65), decreasing biodiversity (0.82), and loss of genetic diversity (0.67). Figure 7 shows the comprehensive indices of the states and impacts on the Lake Malombe ecosystem.

The coefficients of linear regression (R squared) for state and impact were above 0.9, suggesting that these dependent variables are the function of the explanatory variable (time) and that the current state and impact will continue to increase with time.

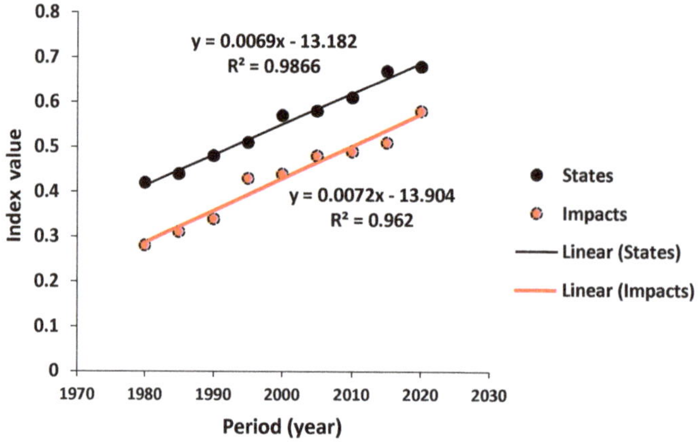

Figure 7. Inter-annual variation of the state and impact index.

3.3. Responses

The response index refers to the measures taken to improve the state of the lake. These measures may act towards any DPSIR indicators within the framework to ensure the lake ecosystem's high efficiency and resilience. Apart from direct actions acting on the DPSIR indicators, the responses focusing on behavioral and attitude changes towards changes to the lake ecosystem through participation and involvement are vital. Different stakeholders interviewed mentioned various strategies to acknowledge the existing drivers, pressures, states, and impacts around Lake Malombe. These strategies were classified into capacity building and governance (0.56), institutional measures (0.36), livelihood intervention (0.62), and environmental interventions (0.59). Capacity building focuses on equipping local communities with the skills and knowledge required to manage the lake ecosystem. These include technical training on biodiversity conservation, ecosystem services, and environmental awareness campaigns.

Governance is focused on the establishment of local structures such as environmental protection committees (EPC), beach village committees (BVC), and fishery associations (FA) to champion the management of the lake ecosystem. Institutional measures include government policies, institutional frameworks, and management strategies. Examples of management policies mentioned during the key informant interviews and FGDs include water, fishery, environmental, climate change, and forestry policies. The legal and institutional frameworks include the Fisheries Conservation and Management Regulations of 2000, the Forestry (Amendment) Act of 2017, the Water Resources (Water pollution control) Regulation Act of 1978 (amended 1997), and the Environment Management Act of 1996 (No. 23 of 1996). The management plans and strategies include the National Biodiversity Strategy and Action Plan II (2015–2025), the National Forest Landscape Restoration Strategy, fisheries management plans such as co-management, and the ecosystem approach to fishery management. Livelihood interventions are activities, such as aquaculture, backyard farming, and livestock production, that aim to promote the local population's welfare. Environmental interventions focus on restoring the lake ecosystem. The environmental interventions identified from the interviews include protecting the fish habitat, conserving and managing the lake catchment (including protected areas such as the gazetted forest in upland areas and national parks' rehabilitation of fish breeding grounds), reafforestation programs in agroforestry, water conservation, and conservation farming. Figure 8 shows the comprehensive indices of responses to drivers, pressures, states, and impacts on the Lake Malombe ecosystem over time.

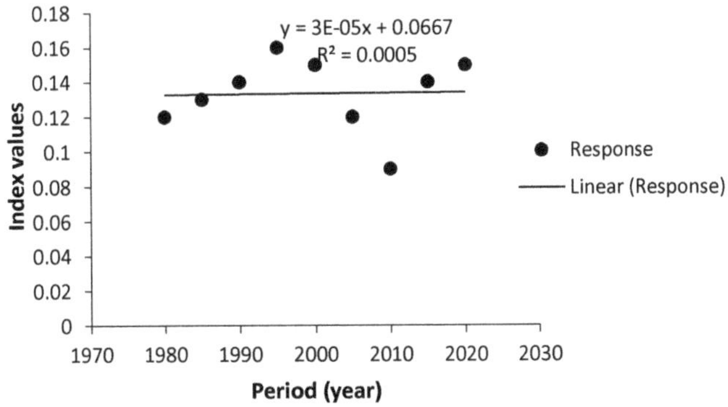

Figure 8. Inter-annual variation of the state and impact index.

The coefficients of linear regression (R squared) for the responses being above 0.0005 suggests that the initiatives taken in Lake Malombe to reverse past and current trends are inadequate to address the drivers, pressures, state, and impacts.

3.4. Lake Malombe Ecosystem Resilience and Influencing Factors

Lake Malombe's ecosystem state can be depicted as a multidimensional surface where troughs and crests represent the lake ecosystem's stable and unstable state. Figure 9 shows that the lake ecosystem is drawn to troughs, in which resilience is equivalent to the depth of a depression that keeps an ecosystem state at equilibrium with time. At the equilibrium, the lake ecosystem is resilient to drivers, pressures, and impacts that noticeably change the ecosystem state around the trough.

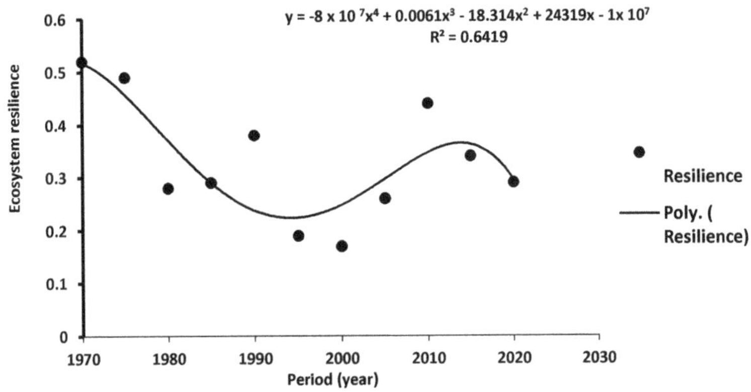

Figure 9. A cup diagram illustrating the Lake Malombe ecosystem's resilience.

A reduction in resilience is illustrated as a shallowing of the depression due to increased demographic, socio–economic, and natural drivers and pressures. Entropy techniques modify the weight of the DPSIR indicators. By calculating the weight of each indicator, key factors that influence the lake ecosystem resilience were identified. The multicollinearity test was conducted before regression analysis to check whether the covariates in the Tobit model were independent of each other. The correlation coefficient matrix in Table 1 showed no multicollinearity in the model, suggesting that the selected variables fit the regression analysis.

Table 1. Correlation matrix of variables.

	Dd	Ds	Dc	P	Ss	Se	I	RCBG	RIM	RLI	REI
Dd	1										
Ds	0.33	1									
Dc	0.12	0.06	1								
P	0.01	0.00	0.03	1							
Ss	0.21	0.31	0.18	0.12	1						
Se	0.02	0.10	0.15	0.09	0.13	1					
I	0.07	0.01	0.21	0.10	0.11	0.20	1				
RCBG	0.02	0.13	0.11	0.13	0.03	0.12	0.03	1			
RIM	0.02	0.12	0.02	0.03	0.02	0.09	0.02	0.07	1		
RLI	0.04	0.06	0.04	0.00	0.02	0.04	0.00	0.08	0.24	1	
REI	0.11	0.1	0.13	0.14	0.11	0.21	0.13	0.16	0.21	0.02	1

Note: Dd = demographic drivers, Ds = socio–economic drivers, Dc = climatic drivers, P = pressures, Ss = socio–economic state, Se = environmental state, I = impact, RCBG = capacity building and governance responses, RIM = institutional measure responses, RLI = livelihood intervention responses, REI = environmental intervention responses.

The Augmented Dickey–Fuller (ADF) co-integration test in Table 2 shows that $p < 0.05$, indicating the existence of a co-integration relationship in the panel data. Demographic, socio–economic, and climatic drivers, pressures, socio–economic state, and environmental state were significant ($p < 0.05$) and negative, suggesting that these indicators negatively influenced the lake ecosystem's resilience.

Table 2. The empirical results of the Tobit model.

DPSIR Indicators	Coefficient	Std.Error	z-Statistic	*p* Value
D_d	−0.060	0.010	−3.201	0.003
D_s	−0.086	0.021	2.710	0.041
D_c	−0.056	0.010	6.619	0.021
P	−0.065	0.010	4.321	0.044
Ss	−0.089	0.007	4.120	0.004
Se	−0.002	0.002	3.231	0.003
I	−0.007	0.001	2.043	0.043
R_{CBG}	0.012	0.012	0.450	0.059
R_{IM}	0.094	0.001	0.013	0.048
R_{LI}	0.063	0.001	−4.234	0.021
R_{EI}	0.012	0.012	6.450	0.012

Heteroskedasticity and autocorrelation-consistent (HAC) ADF *t*-test = −88.032, HAC = 0.003, residue variance = 0.01, *p* value = 0.001. Note: Dd = demographic drivers, Ds = socio–economic drivers, Dc = climatic drivers, P = pressures, Ss = socio–economic state, Se = environmental state, I = impact, RCBG = capacity building and governance responses, RIM = institutional measure responses, RLI = livelihood intervention responses, REI = environmental intervention responses.

On the other hand, response indicators such as livelihood and environmental intervention responses were positive and significant ($p < 0.05$), suggesting that these indicators positively influenced the lake ecosystem resilience.

4. Discussion

4.1. Drivers and Pressures

This study depicts the drivers, pressures, states, impacts, and responses that influence the resilience of inland tropical freshwater shallow lakes' ecosystems. Among the drivers, rapid population growth has resulted in multiple pressures that degrade the state of the Lake Malombe ecosystem. Evidence shows that in 1998 the western bank of Lake Malombe had 8400 households and the eastern bank had 2657 [32]. Using the Malawi national average household size of a five-person per household, the total population in 1998 was estimated at 42,000 on the west bank and 13,285 on the east bank [67]. Factor in 3% annual population growth rate for Mangochi district [38], and the projected population in 2020 was estimated at 82,891 on the west bank and 25,657 households on the east bank. However, the FISH project report revealed that the population trend in the South East Arm where Lake Malombe is located has increased from 295,891 in 1998 to 404,850 in 2008 [27]—making it one of the most densely populated areas (average of over 500 people/km^2) in Malawi. This demographic trend is linked to poverty [33]. The 2011 integrated household survey ranked Mangochi, where Lake Malombe is located, as second among the top four districts (Machinga, Balaka, and Zomba) in Malawi with the highest rates of poverty (0.13 USD/day) and first in terms of the rate of ultra-poor residents (0.08 USD/day)—compared to the national average of 2 USD/day) [33]. Approximately 99% of the local population in the catchment area predominantly depend on fishing for their sustenance [35]. Other sources of income include fish sales from aquaculture, cash crops, rainfed and irrigated food crops, employment, rentals, non-farm businesses, and others. The daily income generated from these socio–economic activities is less than 2 USD/day with the exception of fishery (Figure 10a).

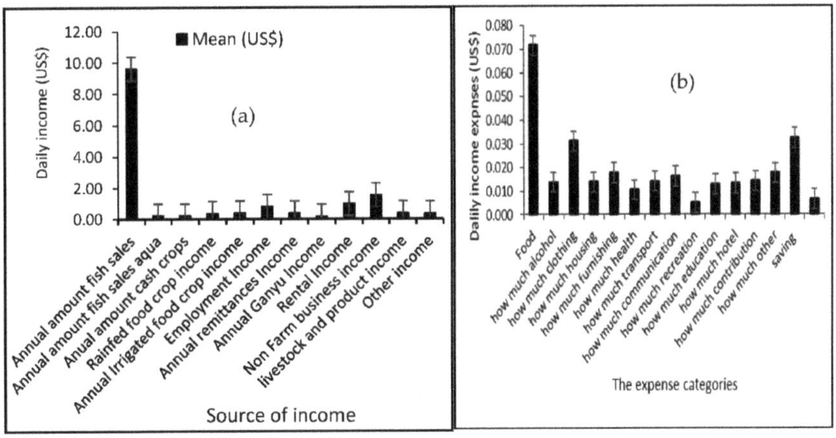

Figure 10. Driving forces of Lake Malombe ecosystem dynamics. (**a**) source of income (**b**) the expense categories.

The expenses in Figure 10b indicate that income is spent mostly on food, while saving is the least common use of money—indicating the local population's socio–economic trap. In the Lake Malombe context, the socio–economic trap is where the local population has limited alternatives to mobilize adequate resources to overcome the shocks and remain in the poverty trap [35,68]. The overall household income is less than 1 USD per day (Figure 10a). The Lake Malawi Basin Program Integrated Household Survey in 2008 put it at 0.74 USD per day, suggesting that the local population's over-dependence on lake resources is triggered by poverty and a lack of alternative livelihood options [52,69]. Sandhu & Sandhu [69] and Suich et al. [70] demonstrated the link between poverty and the ecosystem's resilience. In the Lake Malombe context, poverty pushes vulnerable individuals to desperately over-depend on the lake and exploit even those resources deemed conserved for future generations [35]. This is also the typical scenario in African Great Rift Valley lakes where over-exploitation of lake ecosystem resources is extensive [71], triggering a multitude of direct effects on the riparian populations while undermining the capacity of the lake ecosystem resources to support human livelihoods [71] continuously. The increase in vertical and horizontal effort is associated with poverty, rapid population growth, and a lack of adaptation strategies to the changing ecosystem [31]. The decline in fish biomass landings reported by Makwinja et al. [41] suggests that vertical fishing effort triggered the lake ecosystem dynamics, which eventually leads to over-exploitation of the resource, leading to socio–economic struggles. The scenario depicted in this study is expected in African freshwater ecosystems where poor and economically vulnerable social groups exploit the ecosystems, either as fisherfolks, crew members, or traders [72,73].

Lake Malombe is categorized as one of the areas in Malawi susceptible to climatic variability. The eastern part of the lake is a drought-prone area [32]. Dry spells and floods in the past have had substantial socio–ecological impacts on the lake ecosystem and the local communities in Lake Malombe [32]. The Lake Malombe hydrological regime indicates that it is shaped by extreme weather events such as increased rainfall, prolonged drought, and temperature fluctuation, which consequently threaten socio–ecological components such as livelihood, aquatic vegetation, and biodiversity [74]. Christia et al. [74] and Thanh et al. [75] have pointed out that increased rainfall variability affects water quality, toxic bacteria blooms, and dissolved nutrient concentration, spreading water-borne diseases. Unexpected extreme weather events such as heavy rains, drought, and rising temperatures add more pressure on the existing challenges such as declining ecosystem productivity, soil fertility, water quality degradation, toxic bacteria blooms, and dissolved nutrient concentration, consequently spreading water-borne diseases [75]. The findings align with what Njaya et al. [40], Okpara et al. [76], Lemoalle et al. [77], Likoya [78], Limuwa et al. [79], and Desta [80] reported in Lake Chilwa, the Shire River Basin, and Lake Malawi in Malawi. Cowx and Aya [81], Jamu et al. [31], Wood et al. [82], Dulanya et al. [26], Makwinja et al. [23], Williams et al. [83], and Nkwanda et al. [84] also pointed out the same scenarios in Chia Lagoon, Central Malawi, Lake Chilwa in Southern Malawi, and the Lilongwe River Basin. Guo et al. [85] also demonstrated how human activities such as draining inland freshwater shallow lakes and land use transformation in the basin influenced seasonal flooding in the Poyang Lake basin in China. In West Africa, Li et al. [86] demonstrated how human activities such as landscape alteration in Niger and the Lake Chad basin are strongly linked to changes in hydrological regimes at the watershed scale, leading to a disproportionately large impact on water balance. Elias et al. [83] linked aggressive agricultural expansion to the collapse of the lake ecosystem of the Ethiopia Central Rift Valley. Agricultural development has been identified as putting significant pressure on the mangrove ecosystem [87]. In the Southern Part of the Andaman Islands, about 6 km^2 of mangroves were cleared out during the last three decades (1979–2013) for fishing, severely impacting the mangrove ecosystem and biodiversity. In Malaysia, human encroachment, agriculture, and coastal erosion have been identified as putting pressure on freshwater ecosystems [54].

4.2. State and Impacts

Human and environmental drivers, alongside other pressures occurring across multiple scales, have shaped the current state of the Lake Malombe ecosystem (Figure 11). Increased phytoplankton production, invasion of alien species, water quality degradation, drying up of rivers, over-exploitation of fisheries, increased erosion, shrinking of the lake, and a catastrophic decline in fish stocks, as reported by Hara [88], Dulanya et al. [26], Froese et al. [89], and Makwinja et al. [35], have depressed the lake ecosystem's resilience. The depletion of high trophic level fish species in the 1980s led to what can be described as top-down cascading effects of human food web alterations [37]—a common scenario in many African inland freshwater shallow lakes' ecosystems [90–92].

Note: Photo (a) was taken in the upper Shire River that connects Lake Malawi to Lake Malombe and shows water hyacinth driven towards Lake Malombe. Picture (b) shows the drying river in the west part of Lake Malombe catchment. Picture (c) shows the degraded river banks. Picture (d) shows the shrinking Lake Malombe. All photos except (d), which Emmanuel Chisesa took, were taken by Rodgers Makwinja during field observations.

Figure 11. The current state of the Lake Malombe ecosystem.

The impact indicator in this study is contributed to by severe biodiversity loss, local communities' exposure to vulnerability, population displacement, increased crime rates, a socio–economic trap, HIV/AIDs prevalence, hunger, disease outbreak, decreasing ecosystem provisioning services (EPSs), reduced climate regulation, reduced supporting services, decreasing biodiversity, and a loss of genetic diversity. Increased removal of mangroves, habitat degradation, fishing efforts, extreme lake level decreases, and rivers' drying up [31] are linked to declining fish biomass and loss of revenue, and this has widened the socio–economic gap between the poor and the rich, as demonstrated by the Gini coefficient of 5.6 and an average income of 1.74 USD per household per day [33]. Diseases such as diarrhea, dysentery, and schistosomiasis are common, particularly in the flooded areas of Lake Malombe. Figure 12a shows that water-borne diseases such as diarrhea are the most prominent, followed by dysentery and then schistosomiasis. Figure 12b shows that drought, floods, heavy storms, and dry spells are more prevalent in the three traditional authorities around Lake Malaombe.

Schistosomiasis in Lake Malombe is linked to a declining population of molluscivorous fishes (*Mylochromis* spp., and *Tramatocranus placodon*) due to human and environmental pressures, which enables *Bulinus nyassanus* (the host of human urinary schistosomes) to survive in open waters [93]. Diarrhea and dysentery are highly linked to a decline in forest and shrubland leading to increased seasonal flooding and water pollution. Similar cases are reported in the Lake Chilwa Basin [94], Tonle Sap Lake in Cambodia [95], and Lake Malawi [40]. Makwinja et al. [5] estimated that Lake Malombe lost approximately 8.92 million USD in revenue from fishing within the period from 1982 to 2019 and about 79.83 million USD from 1999 to 2019 due to landscape transformation. In line with these findings, Malawi is currently losing approximately 93 million USD annually due to a decline in ecosystem service functions and unsustainable use of biodiversity [96].

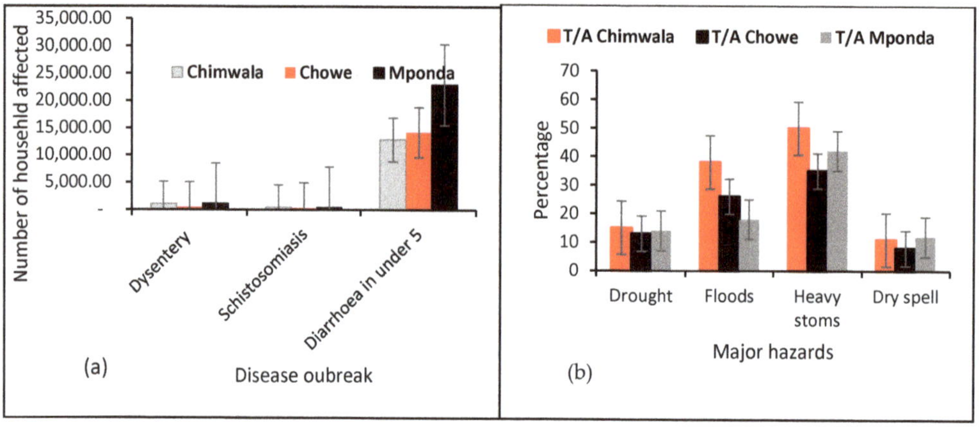

Figure 12. Lake Malombe disease outbreak and significant hazards. (**a**) Disease oubreak. (**b**) Major hazards.

4.3. Response

Response in this study refers to the actions adopted by the decision-makers, managers, and local communities to reverse the situation [97]. Figure 4 categorized the responses into an institutional framework consisting of livelihood intervention and environmental intervention. The institutional framework refers to the government policies and other related legal frameworks formulated to manage freshwater ecosystem service functions in Malawi. The Fisheries Act contains regulations that aim to maximize fishery production with a maximum sustainable yield (MSY) as a reference point. The Fisheries Department, through this act, enforces a closed season, minimum mesh size, licensing, gear restriction, and habit protection in Lake Malombe. However, the challenge is that the act focuses on reducing the vertical and horizontal fishing effort to manage *Oreochromis* species. Other existing threats, such as the declining population of the potamodromous (*O. microlepis*), the commercial extinction of *L. mesops*, the impacts of climate change, and the effect of siltation on fish biodiversity, are given low priority [31]. The command–control nature of the act also makes it costly and challenging to implement due to increased non-compliance and strong resistance from the local fishing communities [98]. Again, Lake Malombe's ecosystem health is shaped by multiple interactions between socio–ecological components rather than a single factor; hence, the Fisheries Act alone cannot effectively reverse the situation. Other related policies such as forestry, water, land, wildlife, irrigation, and tourism formulated to safeguard biodiversity and the lake ecosystem lack synergy and have conflicting key priorities that compromise the conservation effort. National Climate Change Management and fisheries policies do not explicitly explain how the lake ecosystem can withstand the shocks instigated by decreasing water levels due to rainfall fluctuation and drought, which cause the disappearance of aquatic vegetation, reducing fish breeding areas and biodiversity. Illegal exploitation of lake ecosystem resources in Lake Malombe is complex and often driven by economic, social, and instructional factors. For example, increased poverty fuels illegal exploitation of the lake's resources. Limited funding, the lack of a robust governance system, increased corruption, and inadequate monitoring and enforcement leave the conservation managers few options to address the lake ecosystem challenges. Livelihood and environmental intervention refer to the projects aimed at reducing the local population's over-dependence on the lake ecosystem services while stimulating their participation in conservation. In Lake Malombe, promoting aquaculture and income-generating activities such as craft-making and bee-keeping as a substitute for expanding farming around Lake Malombe, charcoal production, and artisanal fishing have been used as livelihood interventions. However, the challenges of these interventions involve the current competing interests among the implementers. Data on the effectiveness and success of these projects have also been scarce, making it difficult to track the progress and their impacts. Environmental intervention refers to those projects focused on ecological restoration and climate change resilience. The success of these interventions depends on how the local population embraces these socio–ecological governance systems and policy reforms. If the communities fail to generate sufficient income from the interventions, they eventually return to lake resource over-exploitation. Figure 8 demonstrates that although the response (R) factor increased slowly, it was the lowest among all indicators, suggesting that the current interventions are inadequate to reverse the situation in Lake Malombe.

4.4. Lake Malombe Ecosystem Resilience and Influencing Factor

Resilience is the ability of the lake to maintain a state of equilibrium after being subjected to external or internal stressors [99]. The freshwater lakes' resilience follows a nonlinear pattern in which perturbations such as drivers, pressures, and impacts move the lakes away from their natural state, but after interventions the balance is restored and the ecosystem returns to its normal state. Lake Malombe, like any other freshwater lake, can maintain its natural ecological state when exposed to drivers, pressures, impacts, and responses. What can be inferred from Figure 9 is that the lake ecosystem resilience is falling away from its natural state due to increasing rates of drivers, pressures, states, and impacts—an indicator of the lake ecosystem's deterioration. Similar scenarios have been reported in China [100], New Zealand [101], and the Mekong river in China, Myanmar, Thailand, Lao PDR, Cambodia, and Vietnam [102]. The effect of demographic, socio–economic, and climatic factors as drivers, pressures, and environmental states were significant ($p < 0.05$) and negative, suggesting that these indicators negatively influenced the lake ecosystem resilience and contributed to its degradation. A similar scenario is depicted in Infranz Wetland, Ethiopia [103] and Dianchi Lake in China [59]. Few response indicators such as livelihood and environmental intervention were positive and significant ($p < 0.05$), suggesting that these indicators slighltly influence the state of the lake ecosystem positively. Although multiple measures have been taken to conserve and protect the lake ecosystem, the results show that there is still much to achieve sustainability and attain the ideal state of the Lake Malombe ecosystem.

As seen in Figures 6 and 7, the steady increase in drivers, pressures, and impacts pushes the lake ecosystem to a deteriorated state. The deteriorated lake ecosystem often shows resilience to restoration activities as strong internal positive feedback loops need to overcome the stressors characterized by internal feedback loops that keep the lake in its state of equilibrium. It has been argued that maintaining the lake ecosystem resilience at a desirable state is the most pragmatic and practical approach to managing the inland freshwater shallow lake ecosystems. Unfortunately, many African freshwater lakes depict the other side of the story. About 40–90% of the shallow lakes in East Africa have dramatically shrunk—the same scenario reported in Lake Chilwa, Lake Chiuta, and other satellite lakes in Malawi [32,38,104]. Figure 9 further shows that the risk index in the Lake Malombe ecosystem reached its worst level between the 1990 and 2000—as evidenced by a lower index value, which indicates the worsening condition of the lake ecosystem. A similar observation was made in New Zealand [101]. Based on the interviews, the responses to the current state of the lake ecosystem in Lake Malombe are hampered by institutional capacities such as lack of funds and training and inadequate technical staff. These limitations are the leading cause of inadequate management plans and strategies for improving the current lake ecosystem's state. Additionally, the continuous removal of forest and grass for farming, fuel, settlements, and construction materials in the upland areas of the catchment will only continue to degrade the lake catchment beds, which are the recipients of debris during the heavy siltation brought by intense precipitation.

5. Conclusions

This study used DPSIR and Tobit models to demonstrate that the state of the Lake Malombe ecosystem is declining, and its vulnerability is linked to demographic, socio–economic, and climatic drivers. It explored the relevant drivers, pressures, states, impacts, and response indicators that are helpful when making management decisions and formulating long-term policy-focused responses. The study further demonstrates that the lake ecosystem resilience is falling away from its natural state due to the increasing rates of drivers, pressures, states, and impacts—an indicator of the lake ecosystem's deterioration and a demonstration of how the top-down processes are gradually eroding the lake ecosystem's health. The study demonstrates that substantial freshwater ecosystem management under the current scenario requires a long-term, resilient sustainability plan. This can be achieved through individual and collective efforts. Medium and long-term conservation efforts and environmental and livelihood projects can help to reduce poverty and over-dependence on the lake ecosystem. The institutional measures for tackling lake resource exploitation should also be adjusted promptly through a series of response measures by local authorities, the local population, and conservation agencies. Challenges related to sedimentation, aquatic biodiversity losses, and water quality degradation require an effective, integrated lake basin restoration programs. There is also a need to integrate different policies relating to land, fisheries, wildlife, climate change, tourism, irrigation, agriculture, food security, and national population to achieve the goal of a sustainable integrated freshwater ecosystem. This integration will help to reduce the existing threats to the lake ecosystem service functions such as cultivation and the conversion of forests, wetlands, and river banks into farmland, which are driven by rapid population growth and poverty.

Author Contributions: R.M. conceptualized the study, developed the methodology, sourced the data, analyzed the data, and developed the original manuscript. S.M., C.C.K., E.K., I.B.M.K., T.A. and F.N. supervised the study, reviewed and edited the manuscript, and visualized and validated the study. All authors have read and agreed to the published version of the manuscript.

Funding: This project was funded by the African Centre of Excellence for Water Management (ACEWM) through the World Bank's African Centres of Excellence (ACE II) Project, grant number GSR/9316/11.

Institutional Review Board Statement: Not applicable.

Informed Consent Statement: Not applicable.

Data Availability Statement: The datasets analyzed during the current study are available from the corresponding author upon reasonable request.

Acknowledgments: The authors wish to thank the Malawi University of Business and Applied Sciences for providing financial support towards the publication of this paper.

Conflicts of Interest: The authors of this paper declare that there are no conflict of interest.

References

1. Reid, A.J.; Carlson, A.K.; Creed, I.F.; Eliason, E.J.; Gell, P.A.; Johnson, P.T.J.; Kidd, K.A.; MacCormack, T.J.; Olden, J.D.; Ormerod, S.J.; et al. Emerging threats and persistent conservation challenges for freshwater biodiversity. *Biol. Rev.* **2019**, *94*, 849–873. [CrossRef] [PubMed]
2. Ho, L.; Goethals, P. Opportunities and Challenges for the Sustainability of Lakes and Reservoirs in Relation to the Sustainable Development Goals (SDGs). *Water* **2019**, *11*, 1462. [CrossRef]
3. Duggan, I.; Rixon, C.; MacIssac, H. Popularity and propagule pressure: Determinants of introduction and establishment of aquarium fish. *Biol. Invasion* **2006**, *8*, 377–382. [CrossRef]
4. Martins, J.; Camanho, M.; Gaspar, M. A review of the application of driving forces-Pressure-State-Impact-Response framework to fisheries management. *Ocean Coast. Manag.* **2012**, *69*, 273–281. [CrossRef]
5. Makwinja, R.; Kaunda, E.; Mengistou, S.; Alamirew, T. Impact of land use/land cover dynamics on ecosystem service value—A case from Lake Malombe, Southern Malawi. *Environ. Monit. Assess.* **2021**, *193*, 1–23. [CrossRef]
6. Fetahi, T. Eutrophication of Ethiopian water bodies: A serious threat to water quality, biodiversity, and public health. *Afr. J. Aquat. Sci.* **2019**, *44*, 303–312. [CrossRef]

7. Olaka, L.A.; Ogutu, J.O.; Said, M.Y.; Oludhe, C. Projected Climatic and Hydrologic Changes to Lake Victoria Basin Rivers under Three RCP Emission Scenarios for 2015–2100 and Impacts on the Water Sector. *Water* **2019**, *11*, 1449. [CrossRef]
8. Ogutu-Ohwayo, R.; Hecky, R.E.; Cohen, A.S.; Kaufman, L. Human impacts on the African Great Lakes. *Environ. Boil. Fishes* **1997**, *50*, 117–131. [CrossRef]
9. Njiru, J.; van der Knaap, M.; Kundu, R.; Nyamweya, C. Lake Victoria fisheries: Outlook and management. *Lakes Reserv.* **2018**, *23*, 52–162. [CrossRef]
10. Dersseh, M.G.; Kibret, A.A.; Tilahun, S.A.; Worqlul, A.W.; Moges, M.A.; Dagnew, D.C.; Abebe, W.B.; Melesse, A.M. Potential of Water Hyacinth Infestation on Lake Tana, Ethiopia: A Prediction Using a GIS-Based Multi-Criteria Technique. *Water* **2019**, *11*, 1921. [CrossRef]
11. IPBES. *Summary for Policymakers of the Global Assessment Report on Biodiversity and Ecosystem Services of the Intergovernmental Science-Policy Platform on Biodiversity and Ecosystem Services*; IPBES: Bonn, Germany, 2019.
12. Earles, J.M.; Yeh, S.; Skog, K.E. Timing of carbon emissions from global forest clearance. *Nat. Clim. Chang.* **2012**, *2*, 682–685. [CrossRef]
13. Makwinja, R.; Mengistou, S.; Kaunda, E.; Alamirew, T. Land use/land cover dynamics, trade-offs and implications on tropical inland shallow lakes' ecosystems' management: Case of Lake Malombe, Malawi. *Sustain. Environ.* **2021**, *7*, 1969139. [CrossRef]
14. Bi, W.; Weng, B.; Yuan, Z.; Ye, M.; Zhang, C.; Zhao, Y.; Yan, D.; Xu, T. Evolution Characteristics of Surface Water Quality Due to Climate Change and LUCC under Scenario Simulations: A Case Study in the Luanhe River Basin. *Int. J. Environ. Res. Public Health* **2018**, *15*, 1724. [CrossRef] [PubMed]
15. Miles, A.; DeLonge, M.S.; Carlisle, L. Triggering a positive research and policy feedback cycle to support a transition to agroecology and sustainable food systems. *Agroecol. Sustain. Food Syst.* **2017**, *41*, 855–879. [CrossRef]
16. Yang, X.-E.; Wu, X.; Hao, H.-L.; He, Z.-L. Mechanisms and assessment of water eutrophication. *J. Zhejiang Univ. Sci. B* **2008**, *9*, 197–209. [CrossRef] [PubMed]
17. Davidson, N.C.; Finlayson, C.M. Extent, regional distribution and changes in area of different classes of wetland. *Mar. Freshw. Res.* **2018**, *69*, 1525. [CrossRef]
18. Makwinja, R.; Kosamu, I.B.M.; Kaonga, C.C. Determinants and Values of Willingness to Pay for Water Quality Improvement: Insights from Chia Lagoon, Malawi. *Sustainability* **2019**, *11*, 4690. [CrossRef]
19. Li, Y.; Zhang, Z.; Shi, M. Restrictive Effects of Water Scarcity on Urban Economic Development in the Beijing-Tianjin-Hebei City Region. *Sustainability* **2019**, *11*, 2452. [CrossRef]
20. Ogutu-Ohwayo, R.; Balirwa, J. Management challenges of freshwater fisheries in Africa. *Lakes Reserv. Res. Manag.* **2006**, *11*, 215–226. [CrossRef]
21. Hall, N.D.; Stuntz, B.B.; Abrams, R.H. Climate Change and Freshwater Resources. *Nat. Resour. Environ.* **2008**, *22*, 30–35.
22. Djoudi, H.; Brockhaus, M.; Locatelli, B. Once there was a lake: Vulnerability to environmental changes in northern Mali. *Reg. Environ. Change* **2013**, *13*, 493–508. [CrossRef]
23. Makwinja, R.; Chapotera, M.; Likongwe, P.; Banda, J.; Chijere, A. Location and Roles of Deep Pools in Likangala River during 2012 Recession Period of Lake Chilwa Basin. *Int. J. Ecol.* **2014**, *2014*, 294683. [CrossRef]
24. Magrin, G. The disappearance of Lake Chad: History of a myth. *J. Political Ecol.* **2016**, *23*, 205–222. [CrossRef]
25. Jones, E.R.; O'Flynn, H.; Njoku, K.; Crosbie, E.J. Detecting endometrial cancer. *Obstet. Gynaecol.* **2021**, *23*, 103–112. [CrossRef]
26. Dulanya, Z.; Croudace, I.; Reed, J.; Trauth, M. Palaeoliminological reconstruction of recent environmental change in Lake Malombe (S. Malawi) using multiple proxies. *Water SA* **2014**, *40*, 717–728. [CrossRef]
27. Okech, E.; Kitaka, N.; Omondi, S.; Verschuren, D. Water level fluctuations in Lake Baringo, Kenya, during the 19th and 20th centuries: Evidence from lake sediments. *Afr. J. Aquat. Sci.* **2019**, *44*, 25–33. [CrossRef]
28. Bootsma, H.; Hecky, R. The major socio-economic benefits from these three major African conservation of the African Great Lakes: A Limnological Perspective. *Conserv. Biol.* **1993**, *7*, 644–656. [CrossRef]
29. Jordaan, M.; Chakona, A. Protected Areas and Endemic Freshwater Fishes of the Cape Fold Ecoregion: Missing the Boat for Fish Conservation? *Front. Environ. Sci.* **2020**, *8*, 502042. [CrossRef]
30. Swallow, B.M.; Sang, J.; Nyabenge, M.; Bundotich, D.K.; Duraiappah, A.K.; Yatich, T.B. Tradeoffs, synergies and traps among ecosystem services in the Lake Victoria basin of East Africa. *Environ. Sci. Policy* **2009**, *12*, 504–519. [CrossRef]
31. Jamu, D.; Banda, M.; Njaya, F.; Hecky, R.E. Challenges to sustainable management of the lakes of Malawi. *J. Great Lakes Res.* **2011**, *37*, 3–14. [CrossRef]
32. FISH. *Environmental Threats and Opportunities Assessment (ETOA) of Four Major Lakes in Malawi*; USAID/FISH Project; Pact Publication: Lilongwe, Malawi, 2015.
33. Makwinja, R.; Kaunda, E.; Mengistou, S.; Alemiew, T.; Njaya, F.; Kosamu, I.B.M.; Kaonga, C.C. Lake Malombe fishing communities' livelihood, vulnerability, and adaptation strategies. *Curr. Res. Environ. Sustain.* **2021**, *3*, 100055. [CrossRef]
34. Makwinja, R.; Mengistou, S.; Kaunda, E.; Alamirew, T. Lake Malombe fish stock fluctuation: Ecosystem and fisherfolks. *Egypt. J. Aquat. Res.* **2021**, *47*, 321–327. [CrossRef]
35. Makwinja, R.; Mengistou, S.; Kaunda, E.; Alamirew, T. Economic value of tropical inland freshwater shallow lakes: Lesson from Lake Malombe, Malawi. *Afr. J. Ecol.* **2021**, *29*, 1–15. [CrossRef]
36. Hara, M.; Njaya, F. Between a rock and a hard place: The need for and challenges to implementation of Rights-Based Fisheries Management in small-scale fisheries of southern Lake Malawi. *Fish. Res.* **2016**, *174*, 10–18.

37. Makwinja, R.; Mengistou, S.; Kaunda, E.; Alamirew, T. Spatial distribution of zooplankton in response to ecological dynamics in tropical shallow lake: Insight from Lake Malombe, Malawi. *J. Freshw. Ecol.* **2021**, *36*, 127–147. [CrossRef]
38. Government of Malawi. *Mangochi District State of Environment and Outlook*; Mangochi District Council: Mangochi, Malawi, 2014.
39. Njaya, F. Ecosystem approach to fisheries in southern Lake Malawi: Status of the fisheries co-management. *Aquat. Ecosyst. Health Manag.* **2018**, *21*, 159–167. [CrossRef]
40. Njaya, F.; Donda, S.; Béné, C. Analysis of Power in Fisheries Co-Management: Experiences from Malawi. *Soc. Nat. Resour.* **2012**, *25*, 652–666. [CrossRef]
41. Makwinja, R.; Mengistou, S.; Kaunda, E.; Alemiew, T.; Phiri, T.; Kosamu, I.; Kaonga, C. Modeling of Lake Malombe Annual Fish Landings and Catch per Unit Effort (CPUE). *Forecasting* **2021**, *3*, 39–55. [CrossRef]
42. Mandić, A. Structuring challenges of sustainable tourism development in protected natural areas with driving force–pressure–state–impact–response (DPSIR) framework. *Environ. Syst. Decis.* **2020**, *40*, 560–576. [CrossRef]
43. Tscherning, K.; Helming, K.; Krippner, B.; Sieber, S.; Paloma, S.G.Y. Does research applying the DPSIR framework support decision making? *Land Use Policy* **2012**, *29*, 102–110. [CrossRef]
44. Rodrigues, J.M.G. Cultural Services in Aquatic Ecosystems. In *Ecosystem Services and River Basin Ecohydrology*; Springer: Dordrecht, The Netherlands, 2015; pp. 35–56.
45. Nguyen, A.T.; Hens, L. Human Ecology of Climate Change Hazards: Concepts, Literature Review, and Methodology. In *Human Ecology of Climate Change Hazards in Vietnam*; Springer: Cham, Switzerland, 2018; pp. 3–36.
46. Ma, X.; Wang, C.; Yu, Y.; Li, Y.; Dong, B.; Zhang, X.; Niu, X.; Yang, Q.; Chen, R.; Li, Y.; et al. Ecological efficiency in China and its influencing factors-a super-efficient SBM metafrontier-Malmquist-Tobit model study. *Environ. Sci. Pollut. Res.* **2018**, *25*, 20880–20898. [CrossRef]
47. Gabrielsen, P.; Bosch, P. *Internal Working Paper Environmental Indicators: Typology and Use in Reporting*; European Environment Agency: Copenhagen, Denmark, 2003.
48. Millennium Ecosystem Assessment. *Ecosystems and Human Well-Being: Synthesis*; Island Press: Washington, DC, USA, 2005.
49. Friend, A.; Rapport, D. Evolution of macro-information systems for sustainable development. *Ecol. Econ.* **1991**, *3*, 59–76. [CrossRef]
50. Borja, Á.; Galparsoro, I.; Solaun, O.; Muxika, I.; Tello, E.M.; Uriarte, A.; Valencia, V. The European Water Framework Directive and the DPSIR, a methodological approach to assess the risk of failing to achieve good ecological status. *Estuar. Coast. Shelf Sci.* **2006**, *66*, 84–96. [CrossRef]
51. Mangi, S.C.; Roberts, C.M.; Rodwell, L.D. Reef fisheries management in Kenya: Preliminary driver–pressure–state–impacts–response (DPSIR). *Ocean Coast. Manag.* **2007**, *50*, 463–480.
52. Bell, R.J.; Collie, J.S.; Jamu, D.; Banda, M. Changes in the biomass of chambo in the southeast arm of Lake Malawi: A stock assessment of *Oreochromis* spp. *J. Great Lakes Res.* **2012**, *38*, 720–729. [CrossRef]
53. Pinto, R.; de Jonge, V.; Neto, J.; Domingos, T.; Marques, J.; Patrício, J. Towards a DPSIR driven integration of ecological value, water uses and ecosystem services for estuarine systems. *Ocean Coast. Manag.* **2013**, *72*, 64–79. [CrossRef]
54. Sarmin, N.; Hasmadi, I.M.; Pakhriazad, H.; Khairil, W. The DPSIR framework for causes analysis of mangrove deforestation in Johor, Malaysia. *Environ. Nanotechnol. Monit. Manag.* **2016**, *6*, 214–218. [CrossRef]
55. Gari, S.R.; Guerrero, C.E.O.; A-Uribe, B.; Icely, J.D.; Newton, A. A DPSIR-analysis of water uses and related water quality issues in the Colombian Alto and Medio Dagua Community Council. *Water Sci.* **2018**, *32*, 318–337. [CrossRef]
56. Zare, F.; Elsawah, S.; Bagheri, A.; Nabavi, E.; Jakeman, A.J. Improved integrated water resource modelling by combining DPSIR and system dynamics conceptual modelling techniques. *J. Environ. Manag.* **2019**, *246*, 27–41. [CrossRef] [PubMed]
57. Gebremedhin, S.; Getahun, A.; Anteneh, W.; Bruneel, S.; Goethals, P. A Drivers-Pressure-State-Impact-Responses Framework to Support the Sustainability of Fish and Fisheries in Lake Tana, Ethiopia. *Sustainability* **2018**, *10*, 2957. [CrossRef]
58. Government of Malawi. *Review of Decentralized Environment Management Guidelines*; Ministry of Local Government and Rural Development: Lilongwe, Malawi, 2011.
59. Wang, Z.; Zhou, J.; Loaiciga, H.; Guo, H.; Hong, S. A DPSIR Model for Ecological Security Assessment through Indicator Screening: A Case Study at Dianchi Lake in China. *PLoS ONE* **2015**, *10*, e0131732. [CrossRef] [PubMed]
60. Niemeijer, D.; de Groot, R. Framing environmental indicators: Moving from causal chains to causal networks. *Environ. Dev. Sustain.* **2008**, *10*, 89–106. [CrossRef]
61. Carr, E.; Wingard, P.M.; Yorty, S.C.; Thompson, M.C.; Jensen, N.K.; Roberson, J. Applying DPSIR to sustainable development. *Int. J. Sustain. Dev. World Ecol.* **2007**, *14*, 543–555. [CrossRef]
62. Heal, G. Valuing ecosystem services. *Ecosystems* **2000**, *3*, 24–30. [CrossRef]
63. Svarstad, H.; Kjeruff, P.; Rothman, D.; Sieple, H.; Watzold, F. Discursive biases of the environmental research framework DPSIR. *Land Use Policy* **2008**, *25*, 116–125.
64. Jury, M.R. Malawi's Shire River Fluctuations and Climate. *J. Hydrometeorol.* **2014**, *15*, 2039–2049. [CrossRef]
65. Zhang, D.; Shen, J.; Sun, F. Evaluation of Water Environment Performance Based on a DPSIR-SBM-Tobid Model. *Water Resour. Hydrol. Eng.* **2020**, *24*, 1641–1654.
66. Amore, M.; Murtinu, S. Tobit models in strategy research: Critical issues and applications. *Glob. Strategy J.* **2021**, *11*, 331–355. [CrossRef]

67. Hara, M.; Jul-Larsen, E. The "Lords" of Malombe; An Analysis of Fishery Development and Changes in Fishing Effort on Lake Malombe, Malaw1. In *Management, Co-Management or no Management? Major Dilemas in Southern African Freshwater Fisheries*; Food and Agriculture Organisation of the United Nations: Rome, Italy, 2003; pp. 179–201.
68. Salas, S.; Bjørkan, M.; Bobadilla, F.; Cabrera, M.A. Addressing Vulnerability: Coping Strategies of Fishing Communities in Yucatan, Mexico. In *Poverty Mosaics: Realities and Prospects in Small-Scale Fisheries*; Springer: Cham, The Switzerland, 2011; pp. 195–220.
69. Sandhu, H.; Sandhu, S. Poverty development, and Himalayan ecosystems. *AMBIO* **2015**, *44*, 297–307. [CrossRef] [PubMed]
70. Suich, H.; Howe, C.; Mace, G. Ecosystem services and poverty alleviation: A review of the empirical links. *Ecosyst. Serv.* **2015**, *12*, 137–147. [CrossRef]
71. Cawthorn, D.; Hoffman, L. The bushmeat and food security nexus: A global account of the contributions, conundrums, and ethical collisions. *Food Res. Int.* **2015**, *76*, 906–925.
72. Welcomme, R.; Cowx, I.; Coates, D.; Béné, C.; Simon, F.; Ashley, H.; Kai, L. Inland capture fisheries. *Philos. Trans. R. Soc. B* **2010**, *365*, 2881–2896. [CrossRef] [PubMed]
73. Nagoli, J.; Mulwafu, W.; Green, E.; Likongwe, P.; Chiwona-Karltun, L. Conflicts over Natural Resource Scarcity in the Aquatic Ecosystem of the Lake Chilwa. *Environ. Ecol. Res.* **2016**, *4*, 207–216. [CrossRef]
74. Christia, C.; Giordani, G.; Papastergiadou, E. Environmental Variability and Macrophyte Assemblages in Coastal Lagoon Types of Western Greece (Mediterranean Sea). *Water* **2018**, *10*, 151. [CrossRef]
75. Thanh, H.T.; Tschakert, P.; Hipsey, M.R. Tracing environmental and livelihood dynamics in a tropical coastal lagoon through the lens of multiple adaptive cycles. *Ecol. Soc.* **2020**, *25*, 31. [CrossRef]
76. Okpara, U.T.; Stringer, L.C.; Dougill, A.J.; Bila, M.D. Conflicts about water in Lake Chad: Are environmental, vulnerability and security issues linked? *Prog. Dev. Stud.* **2015**, *15*, 308–325. [CrossRef]
77. Lemoalle, J.; Bader, J.-C.; Leblanc, M.; Sedick, A. Recent changes in Lake Chad: Observations, simulations and management options (1973–2011). *Glob. Planet. Chang.* **2012**, *80–81*, 247–254. [CrossRef]
78. Likoya, E. Climate Change in the Context of Changing Land Use and Cover: Case Study of the Shire River Basin Flood of 2015. Master's Thesis, University of Cape Town, Cape Town, South Africa, 2019.
79. Limuwa, M.M.; Sitaula, B.K.; Njaya, F.; Storebakken, T. Evaluation of Small-Scale Fishers' Perceptions on Climate Change and Their Coping Strategies: Insights from Lake Malawi. *Climate* **2018**, *6*, 34. [CrossRef]
80. Desta, H.; Lemma, B.; Albert, G.; Stellmacher, T. Degradation of Lake Ziway, Ethiopia: A study of the environmental perceptions of school students. *Lakes Reserv. Res. Manag.* **2015**, *20*, 243–255. [CrossRef]
81. Cowx, I.G.; Aya, M.P. Paradigm shifts in fish conservation: Moving to the ecosystem services concept. *J. Fish. Biol.* **2011**, *79*, 1663–1680. [CrossRef]
82. Wood, A.; Dixon, A.; McCartney, M. *Wetland Management and Sustainable Livelihoods in Africa*; Routledge: London, UK; New York, NY, USA, 2013.
83. Williams, M.; Ryan, C.; Rees, R.; Sambane, E.; Fernando, J.; Grace, J. Carbon sequestration and biodiversity of re-growing miombo woodlands in Mozambique. *For. Ecol. Manag.* **2008**, *254*, 145–155. [CrossRef]
84. Nkwanda, I.S.; Feyisa, G.L.; Zewge, F.; Makwinja, R. Impact of land-use/land-cover dynamics on water quality in the Upper Lilongwe River basin, Malawi. *Int. J. Energy Water Resour.* **2021**, *5*, 193–204. [CrossRef]
85. Guo, H.; Hu, Q.; Jiang, T. Annual and seasonal streamflow responses to climate and land-cover changes in the Poyang Lake basin, China. *J. Hydrol.* **2008**, *355*, 106–122. [CrossRef]
86. Li, K.; Coe, M.; Ramankutty, N.; De Jong, R. Modeling the hydrological impact of land-use change in West Africa. *J. Hydrol.* **2007**, *337*, 258–268. [CrossRef]
87. Dasgupta, R.; Shaw, R. Cumulative Impacts of Human Interventions and Climate Change on Mangrove Ecosystems of South and Southeast Asia: An Overview. *J. Ecosyst.* **2013**, *2013*, 1–15. [CrossRef]
88. Hara, M. Restoring the chambo in Southern Malawi: Learning from the past or reinventing the wheel? *Aquat. Ecosyst. Health* **2016**, *9*, 419–432. [CrossRef]
89. Froese, R.; Walters, C.; Pauly, D.; Winker, H.; Weyl, O.L.F.; Demirel, N.; Tsikliras, A.C.; Holt, S.J. A critique of the balanced harvesting approach to fishing. *ICES J. Mar. Sci.* **2015**, *73*, 1640–1650. [CrossRef]
90. Tweddle, D.; Turner, G. Evidence for lacustrine breeding by sanjika Opsaridium microcephalum (Teleostei: Cyprinidae) in Lake Malawi. *Afr. J. Aquat. Sci.* **2014**, *39*, 479–480. [CrossRef]
91. Weyl, O.; Ribbink, A.J.; Tweddle, D. Lake Malawi: Fishes, fisheries, biodiversity, health and habitat. *Aquat. Ecosyst. Health Manag.* **2010**, *13*, 241–254. [CrossRef]
92. Mkumbo, O.C.; Marshall, B. The Nile perch fishery of Lake Victoria: Current status and management challenges. *Fish. Manag. Ecol.* **2015**, *22*, 56–63. [CrossRef]
93. Stauffer, J.R.; Madsen, H. A one health approach to reducing schistosomiasis transmission in Lake Malawi. *Prev. Med. Community Health* **2018**, *1*, 1–4. [CrossRef]
94. Khonje, A.; Metcalf, C.A.; Diggle, E.; Mlozowa, D.; Jere, C.; Akesson, A.; Corbet, T.; Chimanga, Z. Cholera outbreak in districts around Lake Chilwa, Malawi: Lessons learned. *Malawi Med. J.* **2012**, *24*, 29–33. [PubMed]
95. Derne, B.; Weinstein, P.; Lau, C.L. Wetlands as Sites of Exposure to Water-Borne Infectious Diseases. In *Wetlands and Human Health. Wetlands: Ecology, Conservation, and Management*; Springer: Dordrecht, The Netherlands, 2015; pp. 45–74.

96. Pullanikkatil, D.; Palamuleni, L.; Ruhiiga, T. Assessment of land-use change in Likangala River catchment, Malawi: A remote sensing and DPSIR approach. *Appl. Geogr.* **2016**, *71*, 9–23. [CrossRef]
97. Gian-Reto, W. Community and ecosystem responses to recent climate change. *Philos. Trans. R. Soc. B* **2010**, *365*, 2019–2024. [CrossRef]
98. Etiegni, C.A.; Irvine, K.; Kooy, M. Participatory governance in Lake Victoria (Kenya) fisheries: Whose voices are heard? *Marit. Stud.* **2020**, *19*, 489–507. [CrossRef]
99. Abebe, G. Cash-for-work and food-for-work programs' role in household resilience to food insecurity in southern Ethiopia. *Dev. Pract.* **2020**, *30*, 1068–1081. [CrossRef]
100. Yu, H.; Yang, Z.; Li, B. Sustainability Assessment of Water Resources in Beijing. *Water* **2020**, *12*, 1999. [CrossRef]
101. Mann, T.; Gerwat, W.; Batzer, J.; Eggers, K.; Scherner, C.; Wenck, H.; Stäb, F.; Hearing, V.J.; Röhm, K.H.; Kolbe, L. Inhibition of human tyrosinase requires molecular motifs distinctively different from mushroom tyrosinase. *J. Investig. Dermatol.* **2018**, *138*, 1601–1608. [CrossRef] [PubMed]
102. Le Anh, H.; Schneider, P. A DPSIR Assessment on Ecosystem Services Challenges in the Mekong Delta, Vietnam: Coping with the Impacts of Sand Mining. *Sustainability* **2020**, *12*, 9323. [CrossRef]
103. Eneyew, B.G.; Assefa, W.W. Anthropogenic effect on wetland biodiversity in Lake Tana Region: A case of Infranz Wetland, Northwestern Ethiopia. *Environ. Sustain. Indic.* **2021**, *12*, 100158. [CrossRef]
104. Elias, E.; Seifu, W.; Tesfaye, B.; Girmay, W. Impact of land use/cover changes on lake ecosystem of Ethiopia central rift valley. *Cogent Food Agric.* **2019**, *5*, 1595876. [CrossRef]

water

MDPI

Article

Buffalo Pound Lake—Modelling Water Resource Management Scenarios of a Large Multi-Purpose Prairie Reservoir

Julie Terry [1,*], John-Mark Davies [2,3] and Karl-Erich Lindenschmidt [1]

1 Global Institute for Water Security, School of Environment and Sustainability, University of Saskatchewan, Saskatoon, SK S7N 3H5, Canada; karl-erich.lindenschmidt@usask.ca
2 Water Security Agency, 10-3904 Millar Avenue, Saskatoon, SK S7P 0B1, Canada; john-mark.davies@wsask.ca
3 School of Environment and Sustainability, University of Saskatchewan, Saskatoon, SK S7N 5C8, Canada
* Correspondence: julie.terry@usask.ca

Abstract: Water quality models are an emerging tool in water management to understand and inform decisions related to eutrophication. This study tested flow scenario effects on the water quality of Buffalo Pound Lake—a eutrophic reservoir supplying water for approximately 25% of Saskatchewan's population. The model CE-QUAL-W2 was applied to assess the impact of inter-basin water diversion after the impounded lake received high inflows from local runoff. Three water diversion scenarios were tested: continuous flow, immediate release after nutrient loading increased, and a timed release initiated when water levels returned to normal operating range. Each scenario was tested at three different transfer flow rates. The transfers had a dilution effect but did not affect the timing of the nutrient peaks in the upstream portion of the lake. In the lake's downstream section, nutrients peaked at similar concentrations as the base model, but peaks arrived earlier in the season and attenuated rapidly. Results showed greater variation among scenarios in wet years compared to dry years. Dependent on the timing and quantity of water transferred, some but not all water quality parameters are predicted to improve along with the water diversion flows over the period tested. The results suggest that it is optimal to transfer water while local watershed runoff is minimal.

Keywords: CE-QUAL-W2; water quality model; reservoir management; water transfers

Citation: Terry, J.; Davies, J.-M.; Lindenschmidt, K.-E. Buffalo Pound Lake—Modelling Water Resource Management Scenarios of a Large Multi-Purpose Prairie Reservoir. *Water* **2022**, *14*, 584. https://doi.org/10.3390/w14040584

Academic Editor: Fernando António Leal Pacheco

Received: 18 January 2022
Accepted: 10 February 2022
Published: 15 February 2022

1. Introduction

A variety of management tools are required to effectively understand, predict, and optimize surface water quality. This includes traditional information such as data from well-designed monitoring programs, and increasingly relies on the use of water quality models. Water quality models are critical for assessing situations that cannot be directly measured, to better understand system constrains, and for examining scenarios, including operational changes. Reservoir management has benefitted especially from the growing application of water quality models to the problems of regulated flowthrough and anthropogenic water demands. These models have been applied globally to issues such as selective withdrawal strategies for optimizing water temperatures and quality e.g., [1–4], and water quality simulations e.g., [5–7]. Other uses include modelling nutrient loading from the watershed [8,9], impacts of pisciculture [10], reservoir contaminant scenarios [11], and impacts of changing climate and streamflow on reservoir water quality [12].

Canada has one of the highest rates of water availability per capita per year [13]. Canada also has one of the world's highest per capita water use [14]. The Prairie region contains 80% of Canada's agriculture [15], with agriculture accounting for an 86.5% share of surface water use in the South Saskatchewan River Basin [16]. The climate of the Prairies is highly variable with an extreme temperature range and periodic floods and droughts [15]. The landscape is semi-arid, and many users compete for limited water resources [17]. Population growth and irrigation expansion will place further pressure on Prairie reservoirs showing signs of water stress [18].

Buffalo Pound Lake

Buffalo Pound Lake (BPL) is the water source for approximately 25% of the provincial population of Saskatchewan, Canada. Water demands from BPL include municipal, industrial, and agricultural abstractions, and the lake is an important recreational waterbody. The importance of BPL has resulted in numerous scientific studies over the years e.g., [19–24], and organizational reports from environmental consultancy firms and government agencies. The water quality of BPL is a continuing concern. The lake is situated in an area that experiences high runoff variability associated with its variable climate and is therefore subject to periods of both high and low runoff. BPL is located within a former glacial-melt channel in the Upper Qu'Appelle River Basin. Glacial till on the prairies has abundant nutrients and salts and most prairie lakes are therefore naturally eutrophic [25]. In contrast, BPL's supply reservoir Lake Diefenbaker (LDief) is located in the South Saskatchewan River Basin. LDief receives much of its water from the Rocky Mountains and acts as a nutrient and sediment trap [26,27]. Water sourced from LDief is therefore generally lower in nutrients, salts, organic carbon, and sediment relative to runoff received from BPL's local watershed.

Several recent studies have evaluated options for augmenting the water supply from LDief to BPL to meet increasing water demand from proposed irrigation expansion projects e.g., [28–31]. However, there are physical constraints on how much water can be diverted to BPL. Water levels in the Qu'Appelle River watershed are managed as a whole system that includes multiple downstream lakes—one of which is a lake tributary to the Qu'Appelle that can act as a large surge tank during periods of high water. The diversion of flow into BPL from LDief is also dependent on channel capacity and the amount of upstream runoff from BPL's local watershed in any given year.

Given the complexity of managing the water system, there is a need to better understand how BPL water quality is affected under different inflow scenarios—notably in years with high runoff from its local watershed versus drier years when the majority of water is transferred from LDief. This in turn can form the basis for developing tools and information to help inform decisions on flow management.

Water quality models are an emerging tool for managing water resources on the Prairies. Within the last few years, the first complex water quality models have been applied to BPL as part of a targeted collaborative research strategy. The most recent study, that tested the sensitivity of a BPL CE-QUAL-W2 (W2) model to its boundary data, concluded that flow management strategy may be the most important aspect of water quality management in BPL [32]. The W2 model was calibrated to a historical period due to the empirical data available at the time the model was developed. Regardless, the results clearly implied that any proposed change in flow regime was anticipated to impact BPL's water quality. This model offers a highly relevant platform for assessing whether inter-basin water diversion from LDief, during periods of high local watershed inflows, makes a difference to water quality.

In this paper we use the W2 model to evaluate the effect of different water diversion volumes on the water quality of BPL. Analysis of diversion scenarios provides insight into how flow management in a variable climate environment impacts water quality. This research updates the pre-existing calibrated W2 model, extending the calibration period by including an additional 6.5 years of recent data. The lake now benefits from an expanded monitoring program including water quality profile data, and the reinstatement of an upstream flow gauge. This new data covers a natural period of several wet years, with large watershed runoff events, followed by several relatively dry years where source water shifted to a higher percentage from LDief. The goal of this research was to test whether the updated model provided suitable outputs to assess different water management scenarios. A secondary goal was to undertake scenario assessments using W2. These would determine how different transfer rates would have affected water quality in BPL after the lake received a substantial amount of poor water from local runoff. We anticipated that increasing flows from LDief would result in higher water quality, as defined by lower levels of nutrients in

BPL. The type of modelling approach used in this study can provide an informative tool for investigating this important management question in other aquatic systems.

Section 2 describes the base model set-up and calibration, and details of how the water diversion scenarios were developed. Section 3 presents the results of the base model calibration and results of the scenarios.

2. Methods

2.1. Study Site

BPL is a shallow (mean depth 3.8 m), long (30 km), and narrow (<900 m) impounded lake. BPL is a cold polymictic, wind-driven lake that readily mixes when ice-free. Air temperatures range from an average daily minimum of −17.7 °C in January to an average daily maximum of 26.2 °C in July [33]. On average the lake is ice-covered from November to April, with approximately 30% of annual mean precipitation (365.3 mm) falling as snowfall.

The lake was impounded in 1939 and the dam was most recently upgraded in 2000 [34]. Since the 1950s water to BPL has been supplemented from the South Saskatchewan River Basin. At first, water was pumped but, after construction of Lake Diefenbaker (LDief) in 1967, water has been diverted from LDief via the Qu'Appelle River Dam through the Upper Qu'Appelle River channel to BPL (Figure 1). The main inflows into BPL are through these controlled inter-basin transfers of water. Mean daily discharges at flow gauge 05JG006 (Elbow Diversion), situated 3.3 kms below LDief (Figure 1), ranged from 1.4 m^3/s to 5.9 m^3/s over 20 years (min 0.013 m^3/s/max 14.2 m^3/s; 1999–2019, Water Security Agency hydrometric data). The Upper Qu'Appelle River channel consists of two sections, a channelized section (35 km) and the meandering natural river channel (62 km). Over recent years, the provincial Saskatchewan Water Security Agency (WSA) has worked to improve channel conveyance including deepening sections and undertaking various erosion control projects. Not considering water needs due to the expanded irrigation proposals, the current channel capacity is considered sufficient to meet present and future anticipated water demands.

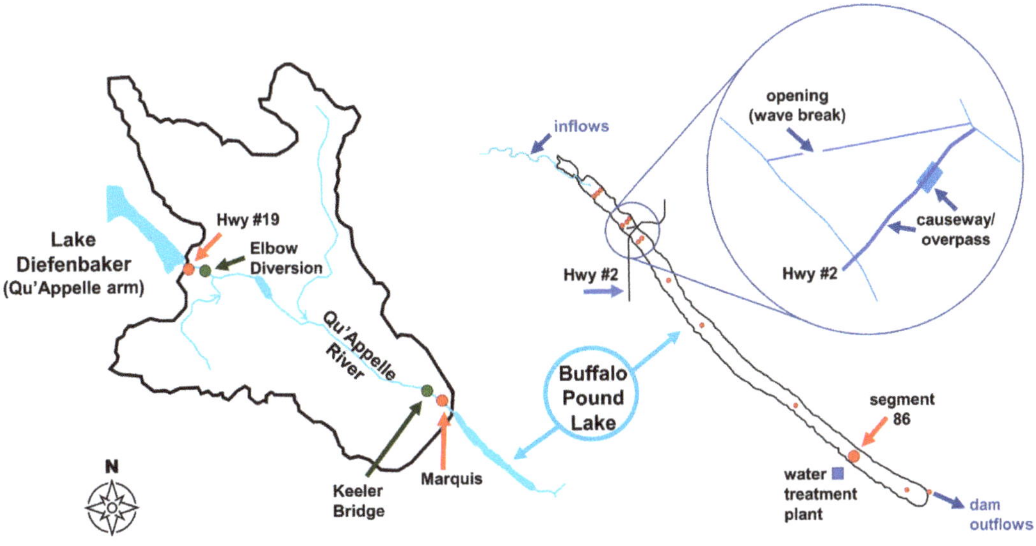

Figure 1. Left: The Upper Qu'Appelle watershed. Right: Buffalo Pound Lake (BPL), Saskatchewan, Canada. The lake has a mean depth of 3.8 m, maximum depth of 6.0 m, average width of 890 m, and total length of 30 kms. Red circles are the water quality sampling stations and green circles are the flow gauges used in this study. Hwy #2 and #19 are official highway numbers.

BPL's water quality is affected by its source waters and in very wet years water quality can be influenced by a surge in nutrients, organic carbon, and dissolved ions from overland run-off [30]. The region has a semi-arid to dry-subhumid climate with long-term ratios of precipitation to potential evapotranspiration (P/PET) of between 0.5 and 0.65 [35,36]. BPL's gross watershed area is 3340 km^2; however, the actual area that contributes flow in any given year is much lower. The area that contributes runoff to BPL in median runoff years, known as the effective drainage area, is 1300 km^2. The reason for the variability of area contributing runoff to BPL includes low P/PET, the generally flat landscape with numerous pothole depressions that are often hydrologically isolated or poorly connected, and differences in landscape water content among years, notably fullness of wetlands and soil moisture content. This means that in dry years BPL receives little runoff from its watershed whereas in wet years it can receive substantial amounts.

The lake is split into separate waterbodies by an old highway (causeway) and the current highway (Figure 1). Inflows to BPL occur in the upper basin. Approximately 4 km from the inflow, the lake is constricted to a 53 m gap at the old highway before entering a triangular lake section between the two highways. The lake then constricts under a 45 m long bridge before entering the main lake body.

Lake residence time is highly variable, ranging from approximately 6 to 36 months [37]. Part of this variability is due to the occasional occurrence of backflows into the lake and variability of watershed runoff among years. When flows in Moose Jaw Creek, a tributary to the Qu'Appelle River downstream of BPL, exceed 50 to 60 m^3/s, water levels increase such that water downstream of BPL flows backwards through the dam into BPL ('backflow'). This occurrence is infrequent (<20% of years) and to minimize the effect on water levels and water quality in BPL the dam's gates are closed; however, water can still backflow into BPL through the dam's fish passage channel and over its radial gates. Typically, water volumes that backflow are low, although in a few years they represent a significant proportion of the lake's volume.

2.2. Model Set-Up

W2 is a complex two-dimensional coupled hydrodynamic and water quality model suitable for use in a range of waterbodies. The model is laterally averaged and includes a dynamic ice model, so is capable of multi-year simulations of a long, narrow, seasonally frozen waterbody such as BPL. The longitudinal segments and vertical layers are user defined and can be variable throughout the grid. Several types of inflow can be specified with individual temperature and constituent files. Numerous hydraulic structures and withdrawals can be placed throughout the grid. The W2 model has been in continuous development since 1986 and has been applied to numerous waterbodies worldwide. A complete description of the model development, hydrodynamic and ecological equations, and model assumptions are provided in the user manual [38]. This study used CE-QUAL-W2 version 4.2.2, the most recent public version at the time of undertaking the model simulations.

A 30 m resolution digital elevation model (DEM) representing the lake's bathymetry from Terry et al. [39] was used in the current project. The bathymetry grid was extended to include the additional 4 km long upper basin (upstream of Highway 2) (Figures 1 and 2). Previous studies did not include this section, making inflows less reliable, as the flow gauge was located above the omitted lake section. This led to uncertainty about the effect that this upper basin had on flows and nutrient transport. The most upstream tip of the lake was too shallow for the boat to navigate when collecting bathymetry and is excluded from the DEM. The shoreline around this section would also complicate the W2 model grid and is above the point of inflows (Figure 1). The DEM's lake extent was created when water levels were close to the operating full supply level, although water levels can occasionally surpass this during very wet years. The W2 model grid extends 1 m vertically from the surface of the DEM to account for these high-water events. The final model grid consists of 100 longitudinal segments around 300 m and up to 28 vertical layers of 0.25 m depth.

W2 has the ability to reflect constrictions in the model's grid, which were included at the highway locations (Figure 2).

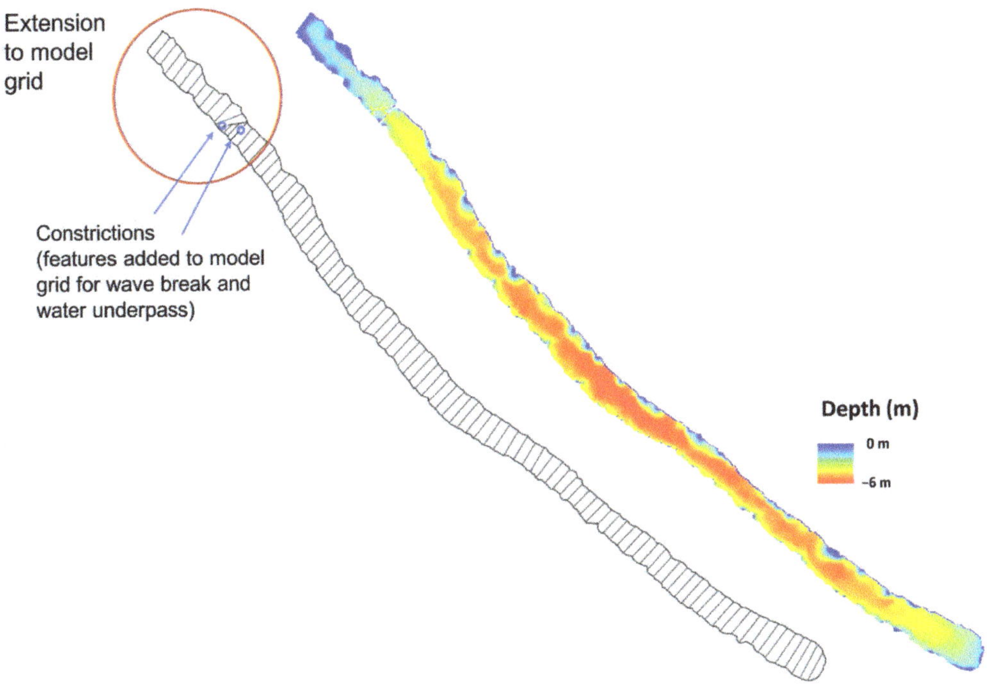

Figure 2. Left: Model grid with longitudinal segmentation. The upstream area in the red circle is the newly extended section of the model. Right: New digital elevation model of BPL including the upper shallow section of lake. Depths are shown relative to operating full supply level.

2.3. Data & Calibration

The W2 simulation period covered 6.5 consecutive years between April 2013 and December 2019. The first years were wet with flood events occurring in 2013–2015. The latter half of the study years were relatively dry.

The hydrological data used in the model were sourced from several datasets. Previous W2 models for BPL had been calibrated using data ending in 1993, after which Environment and Climate Change Canada (ECCC) discontinued the flow gauges above and below BPL. ECCC reinstated the upstream flow gauge (05JG004) in June 2015. Herein, gauge 05JG004 is referred to as the Keeler Bridge gauge (Figure 1). The ECCC downstream gauge has not been reinstated to date; however, the WSA began recording operating logs for the BPL dam structure outflows in January 2014. The WSA conducted a nutrient mass-balance study of the length of the Qu'Appelle River between April 2013 and March 2016 and derived inflows and outflows to BPL [40].

Giving preference to the gauged data, inflows were taken from the WSA nutrient mass-balance study (1 April 2013–31 May 2015), and ECCC flow gauge 05JG004 (1 June 2015–31 December 2019). Outflows from the BPL dam were provided by the WSA nutrient mass-balance study (1 April 2013–31 December 2013), and WSA dam operations log (1 January 2014–31 December 2019). Withdrawal data were provided by the Buffalo Pound Water Treatment Plant (BPWTP) for the main intake pipes to its treatment facility. Withdrawal amounts for a smaller intake pipe supplying surrounding industries were extrapolated from 1995–2013 data, as up-to-date data could not be obtained. Backflows into BPL during a downstream flood event occurred in spring 2013 and spring 2015 and were derived by the WSA. These were included in the model as a tributary entering the most downstream segment. ECCC in-lake water level observations—5-min intervals averaged over 24-h—were used to close the water balance using the W2 water balance tool. This tool is recommended by the developers of W2 and prevents instabilities in the model during simulation due to abrupt changes in water levels. Flows may be positive or negative and are generally added to the model as a distributed tributary entering all segments equally. Adjustment flows were greatest during spring freshet (April 2013, 2014, 2015, 2018) and large precipitation events (June 2014, July 2015) and were assumed to be principally related to ungauged runoff.

Hourly meteorological data (air temperature, dewpoint temperature, wind speed and direction) were downloaded for ECCC climate station Moose Jaw A, located approximately 40 kms south of the lake. The landscape is flat and open between the two locations, and it is expected that the weather conditions will not have differed significantly for modelling purposes. Cloud cover data were not always available and any gaps in cloud cover data at Moose Jaw A were filled in from the nearby ECCC climate station at Regina International Airport. Daily total precipitation was from the ECCC climate station 'Buffalo Pound Lake' located at the site of the BPWTP. Precipitation temperature was set at dewpoint temperature as per W2 recommendations, or zero if the dewpoint was negative. Evaporation was calculated internally by the model based on provided meteorological data and geographic location.

BPL inflow water quality constituent concentrations and water temperatures were from measurements collected at Marquis station (Figure 1). For the two backflow events, backflowing constituent concentrations and water temperatures were taken from measured data at the Buffalo Pound Dam. Precipitation water quality constituent concentrations were not modelled.

The WSA provided water quality measurement data for June 2015–December 2019 for 13 in-lake stations, plus one station at the dam outflow. The 13 in-lake stations were located within eight different model segments. Data were averaged for segments with more than one station if recorded on the same day. The BPWTP also provided weekly observations for April 2013–December 2019 from a long-term dataset. The weekly samples are collected in the early morning at the BPWTP pumping station, located on the downstream end of the lake. Water is pumped from the lake through two pipes approximately one metre off the lakebed. The water travels three kilometers to the pumping station, rising approximately 82 m in elevation. The BPWTP intake pipes are in segment 86 of the model grid. A WSA sampling station is also located in this segment.

The historical BPL WQ model was calibrated in-depth through sensitivity analyses and semi-automated calibration [32,39,41]. For this study, parameter calibration values were initially set based on the older model. With the new data added, the eight model segments corresponding with water quality stations were plotted and results visually compared against the June 2015–December 2019 WSA data. The BPWTP data were plotted in segment 86 to validate the period April 2013–May 2015. Algal taxonomy data did not exist during previous modelling, and in this study the algal rates were updated to better reflect group composition per BPWTP data. Modelled groups were diatoms, green-algae, and cyanobacteria. The organic matter decay rates, and sediment oxygen demand were also recalibrated using the new WSA longitudinal profile water quality data. Wet years and dry years were treated the same in the model. The principal coefficients for the final calibrated model are listed in Table 1. Plots showing model outputs are shown in the results section. Plots focus on model segments 5, located at the top section of the upper basin, segment 32, located near the upstream end of the main lake body downstream of the highway, and segment 86, where the BPWTP intake is located. A description of the ice module and relevant coefficients can be found in Terry et al. [41].

Table 1. Main coefficients used in the water quality model (previous model rates in parentheses if changed) from Terry and Lindenschmidt [32]).

Coefficient	Description	Value		Units
TSED	Sediment temperature	10.3		°C
CBHE	Coefficient of bottom heat exchange	0.3		$W\ m^{-2}$
SOD	Zero-order sediment oxygen demand	0.1–1.2 (1.2)		$g\ O_2\ m^{-2}\ day^{-1}$
PO4R	Sediment release rate of phosphorus (fraction of SOD)	0.001		
NH4R	Sediment release rate of ammonium (fraction of SOD)	0.001		-
NH4DK	Ammonium decay rate	0.12		day^{-1}
NO3DK	Nitrate decay rate	0.1		day^{-1}
LDOMDK	Labile DOM decay rate	0.1 (0.25)		day^{-1}
RDOMDK	Refractory DOM decay rate	0.001 (0.012)		day^{-1}
LRDDK	Labile to refractory DOM decay rate	0.01 (0.001)		day^{-1}
LPOMDK	Labile POM decay rate	0.08 (0.32)		day^{-1}
RPOMDK	Refractory POM decay rate	0.001 (0.012)		day^{-1}
LRPDK	Labile to refractory POM decay rate	0.01		day^{-1}
SSS	Suspended solids settling rate	1.0		$m\ day^{-1}$
EXSS	Extinction due to inorganic suspended solids	0.01		$m^{-1}/(g\ m^{-3})$
EXOM	Extinction due to organic suspended solids	0.01		$m^{-1}/(g\ m^{-3})$
EXH20	Light extinction coefficient for pure water	0.25		m^{-1}
BETA	Fraction of incident solar radiation absorbed at the water surface	0.55		-
WSC	Wind shelter coefficient	0.9		°C
	Algal Coefficients	**Diatoms**	**Greens**	**Cyanobacteria**
AG	Maximum algal growth rate, day^{-1}	1.5 (2.5)	2.0 (1.0)	0.5 (0.9)
AM	Maximum algal mortality rate, day^{-1}	0.1	0.1 (0.15)	0.1
AS	Algal settling rate, $m\ day^{-1}$	0.2 (0.02)	0.1 (0.15)	0.02 (0.1)
AHSP	Algal half-saturation for phosphorus limited growth, $g\ m^{-3}$	0.003	0.003	0.003
AHSN	Algal half-saturation for nitrogen limited growth, $g\ m^{-3}$	0.014	0.014	0 * (0.01)
AT1	Lower temperature for algal growth, °C	2.0	10.0	10.0
AT2	Lower temperature for maximum algal growth, °C	8.0	30.0	35.0
AT3	Upper temperature for maximum algal growth, °C	15.0	35.0	40.0
AT4	Upper temperature for algal growth, °C	24.0	40.0	50.0
ACHLA	Ratio between algal biomass and chlorophyll a in terms of mg algae/μg chl a	0.05	0.04	0.1

* Denotes nitrogen fixation–Dolichospermum spp. (formerly Anabaena) are one of the dominant cyanobacteria taxa in BPL.

2.4. Scenarios

Three scenarios were developed for modelling LDief releases, with each scenario being tested at flow rates of 4 m^3/s, 8 m^3/s, and 12 m^3/s for a total of nine simulations. These rates are within the hydrological constraints of the current Upper Qu'Appelle system. Scenario one is the constant flow scenario. It simulates a constant release rate from LDief for the entire model period. This would be a situation where the augmented water transfers from LDief occurred on a permanent year-round basis. There may be practical constraints to transferring water in wet years; however, the simulation is of theoretical interest.

Scenario two is the immediate release scenario. This was designed to investigate BPL's response if LDief releases started immediately after the first high flow event occurred in 2014. Vandergucht et al.'s [40] nutrient mass-balance report found that inflows at Keeler Bridge and nutrient concentrations at Marquis increased rapidly on 7 April 2014. Flows peaked at 65.6 m^3/s on 9 April 2014. Model simulated releases commenced as soon as flows returned to rates <10 m^3/s on 16 April 2014. This scenario was designed to mimic a management response of transferring additional water from LDief in reaction to increased nutrient loading into BPL. Modelled transfers started when flow in the Upper Qu'Appelle subsided sufficiently in consideration of channel capacity and continued until the end of the simulation. As with the previous simulation there are practical limitations with this approach during subsequent high-flow events, but this scenario specifically focusses on the response of BPL on initiating LDief transfers.

Scenario three commences after the first 2014 high flow event but with a delayed initiation and for a defined duration, referred to as the timed release scenario. Water transfers from LDief started on 9 May 2014. This date coincides with the time when 2014 flood levels in BPL lowered to near operating full supply level. This scenario assumed that additional water from LDief could not be added to BPL until water levels were sufficiently low. Water transfers terminated on 12 March 2015, when the following spring freshet caused notable increases in flow.

The scenarios were included in the model structure as a tributary entering the same upstream segment as the main inflows, which reflects the nature of the system's hydrology. For each simulation, outflows of equal rates were added to the BPL dam outflow file to aid the water balance. All scenarios used the same tributary water temperature and constituent files.

Scenario Input Files

Flow, water temperature, and constituent concentrations were required for input files for simulating the water releases from LDief. Water temperatures were assumed to be the same as the temperatures observed at Marquis station. Although temperature data existed for Highway 19 (Hwy 19), located 1.9 kms of channel length below LDief's outflow (Figure 1), consistent temperature changes occurred along the length of the Upper Qu'Appelle channel. Data at the Marquis station were more representative of the temperature of water flowing into BPL regardless of the source.

As with water temperature, constituent concentrations change along the Upper Qu'Appelle channel. Nutrients in rivers and streams cycle between the water column and sediments during transit [42,43]. Sources and sinks include recycling by aquatic biota, settling and resuspension of the sediment bed, adsorption to suspended solids, and biochemical transformations such as denitrification. Reoccurring patterns in constituent changes from upstream to downstream were examined so that they could be accounted for in the simulated LDief releases. Constituent concentrations at both Hwy 19 and Marquis water quality stations were compared during periods when flows were similar at both stations, a time when overland flow and tributary contributions from the watershed would have been minimal. Transit time, as a function of flow at Elbow Diversion (Figure 1), was calculated using a logarithmic function Equation (1) produced by a one-dimensional hydrologic river model (HEC-RAS) applied previously to the Upper Qu'Appelle River for winter flow testing [44]:

$$\text{Transit Time (days)} = -1.178 * \ln(\text{flow}) + 4.7491; (R^2 = 0.9742) \quad (1)$$

Using the transit time equation, the flow at Elbow Diversion and water quality at Hwy 19 were matched with their downstream counterparts at Keeler Bridge and Marquis. Where the absolute difference in flow was greater than 0.5 m^3/s the corresponding pair of water quality observations were removed from the sample set. For the remaining pairs of water quality observations, changes in concentration were calculated. Ordinary least squares regression analysis was performed between these differences and flow at Elbow Diversion to test if the regression model could be used to estimate net change in LDief concentrations for a given flow rate. The analysis was restricted to observations after 1 June 2015, which was the date the Keeler Bridge flow gauge was reinstated. In April 2016 the WSA changed their water quality reporting laboratory. For orthophosphate (PO_4^{3-}-P), ammonium (NH_4-N), and nitrate-nitrite (NO_3-N + NO_2-N, or NOx), many observations after this date were at or under new reporting limits of detection of PO_4^{3-}-P < 0.02 mg/L, NH_4-N < 0.02 mg/L, and NOx < 0.04 mg/L. The change in concentrations between the two water quality stations could not be calculated, and these observations were removed from the paired samples for the regression analysis. A greater number of samples may have returned different relationships between the three constituents and flow.

The difference between upstream and downstream concentrations for total suspended solids (TSS), total phosphorus (TP), PO_4^{3-}-P, dissolved organic carbon (DOC), and total dissolved solids (TDS) were found to be significantly related to flow (Table 2); however, for DOC and TDS, the amount of variance explained was low. The strongest relationships were for TSS and phosphorus (P).

Table 2. Regression models of the net change in concentration values for constituents travelling along the Upper Qu'Appelle River, as a function of flow at the upstream Elbow Diversion gauging station. Constituents are total suspended solids (TSS), total phosphorus (TP), orthophosphate (PO_4^{3-}-P), dissolved organic carbon (DOC), total dissolved solids (TDS), ammonium (NH_4-N), total nitrogen (TN) and nitrate-nitrite (NOx). Flow is average daily flow in m^3/s and constituents are in mg/L.

	Observations	Adjusted R Square	Standard Error	*p*-Value (Flow)	*p*-Value (Intercept)
TSS (TSS = 24.600 ∗ Flow − 3.880)	40	0.567	61.756	<0.001	0.808
TP (ΔTP = 0.025 ∗ Flow − 0.008)	40	0.540	0.067	<0.001	0.655
PO_4^{3-}-P (ΔPO_4^{3-}-P = 0.007 ∗ Flow − 0.002)	*8*	*0.511*	*0.019*	*<0.05*	*0.828*
DOC (ΔDOC = −0.262 ∗ Flow + 2.064)	40	0.133	1.798	<0.05	<0.001
TDS (ΔTDS = −42.700 ∗ Flow + 331.88)	40	0.123	303.886	<0.05	<0.001
NH_4-N (ΔNH_4-N = −0.389 ∗ Flow + 3.256)	8	0.051	2.722	0.286	0.080
TN (ΔTN = 0.023 ∗ Flow + 0.003)	40	0.032	0.278	0.138	0.969
NO_x (ΔNO_x = −0.002 ∗ Flow − 0.01)	23	−0.030	0.046	0.551	0.610

The change in concentrations for TSS, TP, and PO_4^{3-}-P are plotted in Figure 3. The regressions show a positive significant relationship with flow. The intercepts of all three relationships were rejected at 5% significance. Soluble P is strongly sorbed to suspended solids and bed sediments [45,46]. As a result, both soluble and particulate P concentrations can be influenced by the same hydrological processes that resuspend and transport sediments at different flow rates. Changes in TP and PO_4^{3-}-P concentrations show similar outcomes to TSS in the regression models. At higher flow rates, linear relationships between suspended solids and flow can break down as mobilized sediments are depleted and increasing water volume dilutes concentrations [47]. This is evident in the pattern for TSS in Figure 3.

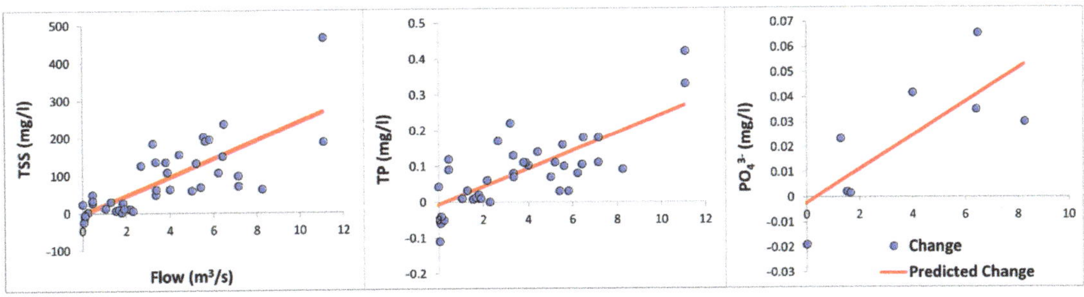

Figure 3. Regression plots for the three constituents that displayed a stronger relationship (adjusted R square > 0.5) between change in concentrations during transport along the Upper Qu'Appelle River with flow rate. Flow is average daily flow in m^3/s at the WSA gauging station Elbow Diversion.

The change in concentrations for variables DOC and TDS had a significant relationship with flow, as seen by the *p*-values. However, for a physical process, the ability of the regression model to explain the variability was low based on the adjusted R square value, and not suitable for predicting the nutrient transformations for the LDief releases. TDS formation and composition is influenced by site specific biological and chemical processes such as decay of organic materials, pH, temperature, and mineral types [48]. DOC chemical composition can be altered through prolonged reservoir storage before downstream release [49], and can be highly variable during storm events [50]. Thus, the transformations and recycling occurring along a river stretch will be inconsistent.

NH_4-N, total nitrogen (TN), and NOx did not have a significant flow relationship. Poor or unexpected relationships between concentrations of particulate and dissolved forms of N with flow have been found previously [51].

The positive relationships for TP and TSS could not be incorporated into the scenario input files. The inflow constituent files in W2 require a split between organic and inorganic substances for the boundary data and the model requires inorganic suspended solids (ISS) rather than TSS. ISS were not included in the modelled constituents as observations just for the inorganic forms were not available. Likewise, variables such as Chlorophyll-a (CHLA), TP, TN, and DOC are not accepted by W2 as inputs but are derived by the model using inflow concentrations for PO_4^{3-}-P, NH_4-N, NOx, algae concentrations, and organic matter (OM). The OM concentrations need to be split into four compartments in the inflow files (labile and refractory components of dissolved and particulate organic matter: LDOM, RDOM, LPOM & RPOM). The organic N, P and carbon content of the OM is calculated internally by W2 based on user-specified fixed ratios. If the organic N and P fractions are known then they can be directly included in the timeseries data as OM sub-compartments (e.g., LDOM-P, LDOM-N, RDOM-P, etc.). W2 does not double count the OM and OM-P/OM-N loading when added in this manner. Available laboratory results for OM did not meet the required W2 inputs. Available data included DOC, and these were used as the basis for estimating OM compartments based on knowledge of the source water.

The positive transformation relationship for PO_4^{3-}-P could be accounted for in the scenario input files, and W2 assumes all phosphorus to be completely available for uptake in this form. The tributary inflow constituent file included PO_4^{3-}-P concentrations after the regression model in Table 2 had been applied. For the remaining inflow constituents in the LDief files, TDS, NH4-N, NOx, LDOM, RDOM, LPOM & RPOM remained at Hwy 19 concentrations, while CHLA and DO used input concentrations from Marquis. No CHLA data were available for Hwy 19, and for DO we use the same assumptions as for water temperature.

3. Results and Discussion

3.1. Model Calibration

Results from the updated model generally correspond well with measured data for the parameters tested (Figure 4). Modelled TDS matched observed data well. This was expected because of the relatively conservative nature of TDS. The variability of inflow TDS within and among years is reflected in the segment 5 results but further down the lake this is muted due to travel length and mixing, as expected. Total DOC followed a similar pattern to TDS. Modelled DO concentrations were largely driven by seasonal changes in temperature. Measured and modelled nutrient concentrations did not match as well as some other parameters, notably in segment 32 for TN, and during the wet years in downstream segment 86. CHLA in segment 5 was modelled well, and the measured CHLA peaks in 2018 and 2019 were reflected by the model. CHLA measurements can be highly variable spatially and temporally, and water quality models should capture the seasonal range of values rather than the variability of individual observations. By segment 86, modelled CHLA had a consistent seasonal pattern, although the amplitude did not capture the range in measured values in all years. In particular, the model did not capture winter algal blooms in the downstream segments. Interestingly, the model results showed an early spring algal bloom occurring in segment 86 in April 2013. This was the timing of the first backflow event into BPL that brought elevated concentrations of nutrients as seen by the corresponding spike in TP and DOC at this time. TDS concentrations were lower in the backflowing water, and the dilution of in-lake concentrations can also be seen in the plots. The second backflow event occurred in March 2015, which can again be seen on the plots of segment 86.

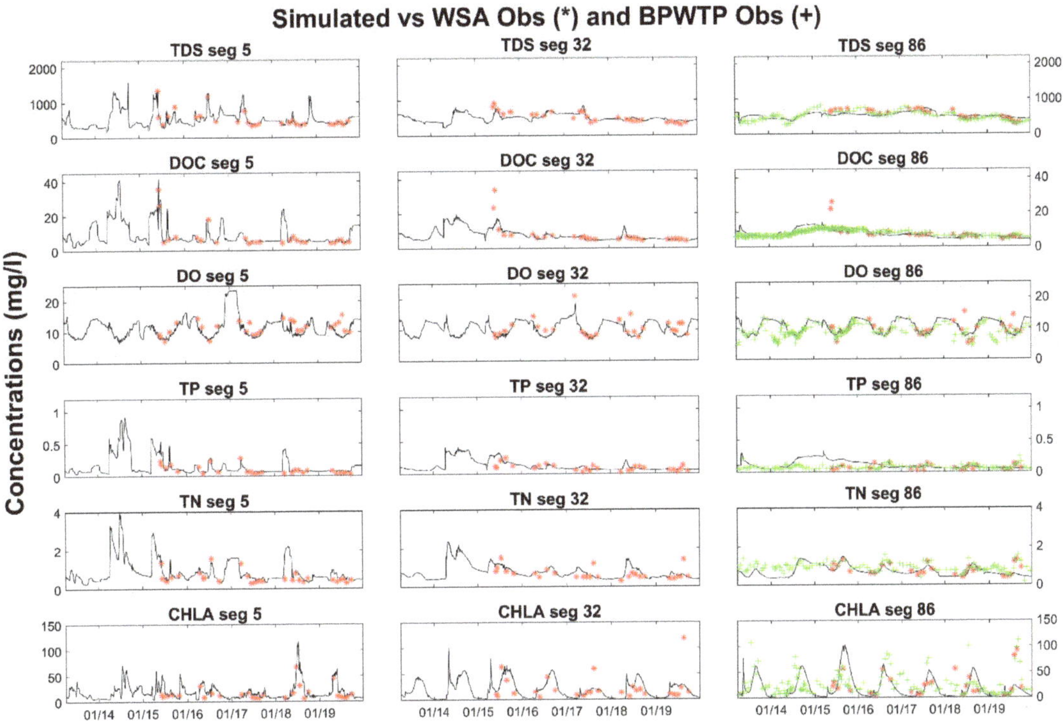

Figure 4. Model results for BPL for the calibration period April 2013–December 2019. Segment 5 is located upstream in the top section of lake. Segments 32 and 86 are both located in the main body of the lake. Water Security Agency observed in-lake sample data (red *) are included from June 2015. Observed weekly data provided by the Buffalo Pound Water Treatment Plant (green +) are shown in segment 86. Inflow constituents DO = dissolved oxygen, and CHLA = Chlorophyll-a.

Spring freshet plus intense rain-events led to several high-flow and loading events in the spring and summer of 2014. High concentrations of TP, TN, and DOC can be seen in the plots for segment 5. TP, TN, and DOC are derived internally in W2, and useful for checking the model's overall treatment of the different sources and sinks of each key constituent. It appears that the nutrient loading in this period was overstated in the model's boundary data. The sampling frequency for the Marquis station water quality data ranged from 3 days to 1 month between April and September 2014, with high concentrations of PO_4^{3-}-P, for example, recorded in all but two of 14 samples. Stored nutrients in the watershed, riverbed, and riverbanks can be quickly mobilized once flows pass a critical threshold. This local mobilization begins on the rising limb of the hydrograph and often peaks in advance of the flow peak [52,53]. More distant nutrient sources in overland runoff and subsurface flow can arrive both before or after the flow peak dependent on the hydrological event and site [52,53]. Travel time in the Upper Qu'Appelle channel can be under two days in high flows, and nutrient spikes dissipate in BPL rapidly. The recorded nutrient concentrations around this time were unlikely to be sustained at high concentrations over the entire five months but, rather, capture several peak events over these wet months. WSA in-lake data were not available for verification of inflowing nutrient concentrations, and concentrations were therefore added as per the available Marquis observations. The

BPWTP weekly sample set indicated that the model was overstating TP, TN, and DOC as far downstream as segment 86 in that year.

The most significant outcome from the development of the BPL model is that the model appeared to handle both extended wet and dry conditions without predicted results deviating too far from actual observations. A major challenge in multi-year simulations is having only one set of coefficients to represent a range of environmental conditions. The model has been calibrated to relatively dry periods, including the original 1986–1993 period. The same coefficients were then used for model validation of two extremely wet years (April 2013–June 2015) with reasonable success, per the segment 86 comparison with the BPWTP data. Benefits of updating the model to the most recent data available were that the water quality scenarios became more relevant to current lake conditions. In addition, the extension of the model grid to include the top section of the lake, and the addition of the new longitudinal profile data means much of the uncertainty was removed from the previous modelling work of Terry and Lindenschmidt [32]. While the parameters of the older model did not require much change, the comparison of output in several segments at once allowed the model assumptions to be evaluated in greater detail. The final model was deemed suitable for running the management scenarios.

3.2. Model Simulations

For all scenarios there is a general pattern change from the wet period, shortly after the start of the scenarios in the spring of 2014, to the dry period after around 2016. The wet period saw larger magnitude concentration changes and larger differences among the scenarios, whereas scenario results were more similar to each other during the dry period (Figure 5).

For the constant flow scenario (Figure 5), of note are the peak concentrations during high flow events. At the upstream end, segment 5 was closely aligned with inflows and inflow concentrations, and the additional constant discharge had a dilution effect but did not affect the timing of the peaks. In segments 32 and 86, the scenario constituents regularly peaked at similar concentrations to the base model, yet the peaks occurred much earlier in the season and decreased more rapidly. The peak events also subsided after each freshet or high precipitation event and multiple peaks of shorter duration were evident. The maximum residence time according to W2 is 338.4 days for the baseline model. This reduces to only 82.2 days when a constant discharge of 12 m^3/s is simulated. The lake is shallow, and materials easily suspend in the water column through wind stress and water flow. The additional discharge velocity from the LDief releases transported the constituents more rapidly and further through the lake model.

Outside of the peak events, TDS, DOC and TN concentrations decreased along with additional LDief flows. TN showed substantial reduction in concentrations in the later drier years at all three flow rates. TDS and DOC also showed reduction in concentrations in these same years, although results were closer to baseline.

TP demonstrated the opposite behaviour, and outside of the large 2014 and 2015 loading events, TP concentrations increased from baseline with each rise in LDief discharge. CHLA largely followed the trend of the nutrients with reduced concentrations in the wet years, when nutrients were smaller than the base model, and higher concentrations in the dry years when TP was increased by the scenarios.

The difference in behaviour of TP can be explained by the PO_4^{3-}-P boundary data. The percentage of PO_4^{3-}-P in TP according to the WSA data between April 2013–March 2016 was 4–100% at Hwy 19 (avg. 32%), and 5–86% (avg. 39%) at Marquis. The W2 model assumes all available P for uptake is of this form with PO_4^{3-}-P a large proportion of derived TP. PO_4^{3-}-P is the only constituent to which a regression model transformation was applied to estimate nutrient transformations in the Upper Qu'Appelle River. The relationship with flow was significant; however, the regression equation was not strong and based on only eight observation pairs, though the regression relationship with TP and flow was also positive (Figure 3), implying a transformation equation of some kind was required. Conversely, if linear relationships break down at higher discharge, as per TSS, then the equation may not have applied equally at all three flow rates across time. Clearly, further work is needed to establish a method for estimating nutrient transformations along the Upper Qu'Appelle. This could be achieved through targeted sampling at different times of year, and/or model chaining or extension of the W2 lake model through the Upper Qu'Appelle channel.

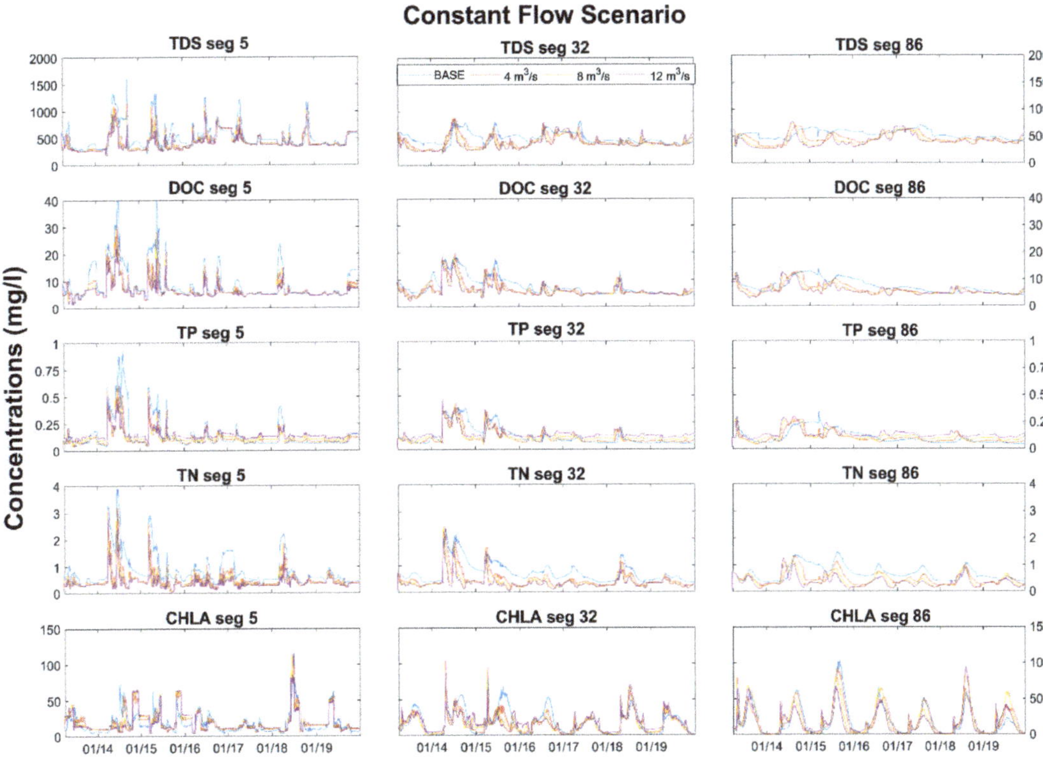

Figure 5. Output for five model-derived water quality constituents of interest at segments 5, 32 and 86 for the constant flow scenario. Scenario flow rates for the water transfers are 4, 8 and 12 m^3/s.

For the immediate release scenario (Figure 6), the model responded quickly to the initiation of the LDief transfers. Concentrations in segment 5 immediately returned to the concentrations modelled in the constant flow scenario, as would be expected so close to the inflow point. In segment 32 concentrations were similar, although TN peaked at slightly higher concentrations in this scenario when the transfers started—most obviously at the transfer rate of 12 m^3/s. By segment 86 there were no noticeable differences in concentration between the scenarios on initiation of transfers. As per the constant flow scenario, constituent peaks decreased in segment 5 with the LDief transfers during the period of high overland runoff. This led to lower concentrations overall in the downstream segments as the model traced the water quality signal from the releases.

For the timed release scenario (Figure 7), transfers were terminated in March 2015 prior to spring freshet. The model responded similarly to the other scenarios on commencement of the water transfers. Of most interest in this scenario was the impact that terminating the transfers ahead of the freshet had on 2015 concentrations. By pre-lowering the concentrations through the LDief transfers, and then reducing overall loading during the freshet itself by stopping the transfers, TDS, DOC, TP, and TN all peaked at far lower concentrations that year in segment 86. Constituent concentrations did not return to higher concentration baseline levels until mid-July 2015, four-months after LDief water transfers were terminated. Segment 86 is of interest as model results are more removed from input values, and it is located at BPWTP intake, so is of most management relevance. It should be noted that concentrations returned to baseline levels much more rapidly in the upstream segments and constituents peaked at similar concentrations to the base model for the 2015 freshet. The timed release is most likely to have returned better results than the constant flow and immediate release scenarios in segment 86 that year as the reduced flow velocity, from the termination of transfers, increased residence time and allowed more materials to settle in the model.

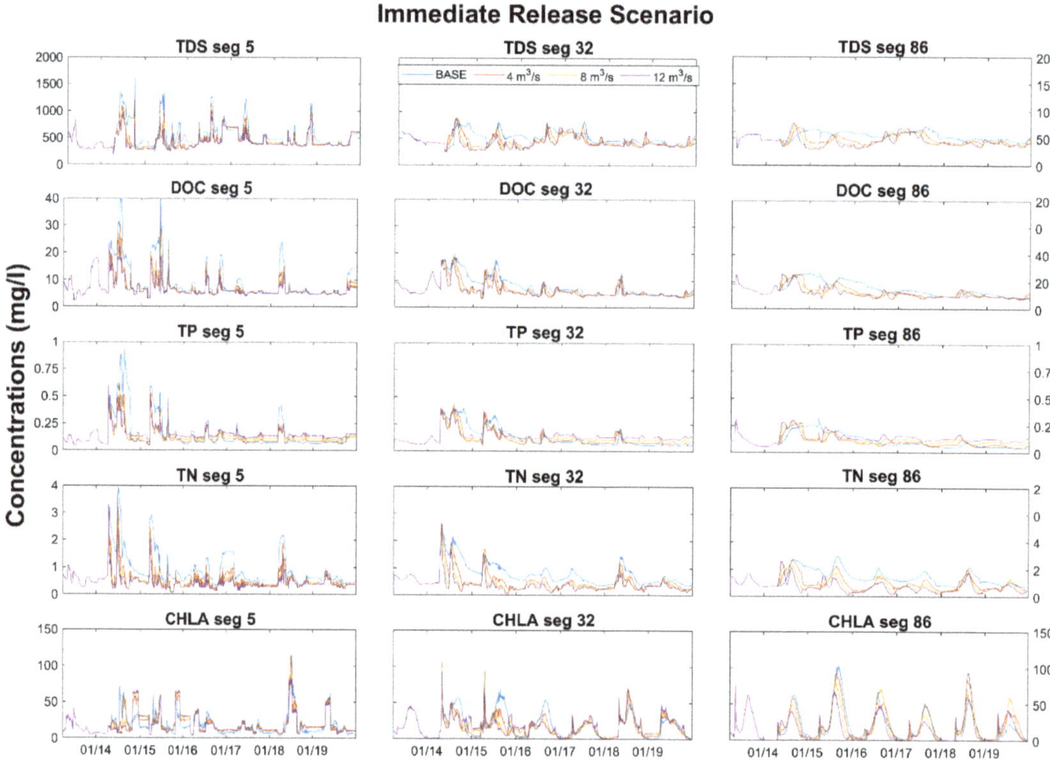

Figure 6. Output for five model-derived water quality constituents of interest at segments 5, 32 and 86 for the immediate release scenario.

The total spread of predicted concentrations at segment 86 illustrates the overall impact on water quality concentrations per scenario (Figure 8). For DOC, TP, and TN results for all scenarios and the base model appear positively skewed. This data distribution infers that most predicted concentrations were small (the box plus left tail includes 75% of the predicted concentrations). The whiskers in Figure 8 are plotted at interquartile range (IQR) $*$ 1.5, with the remaining data, beyond the whiskers, representing the infrequent high concentration loading events (e.g., freshet). TDS demonstrated increasing skewness with the duration of LDief releases and increased transfer rates.

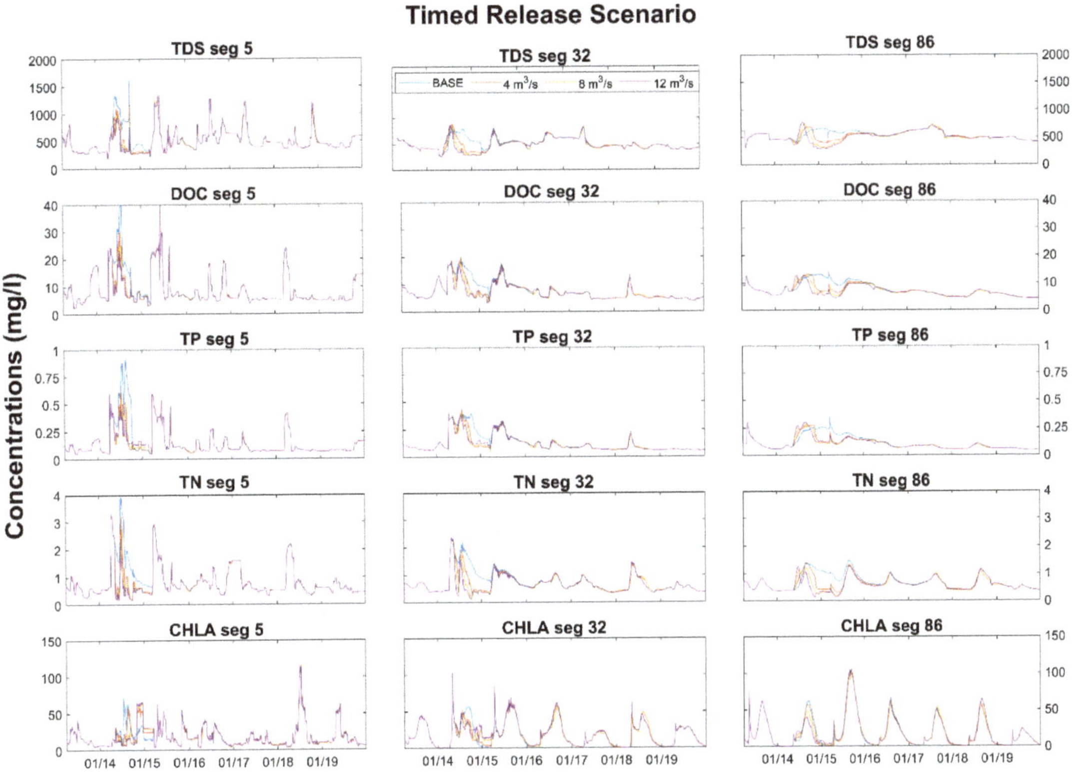

Figure 7. Output for five model-derived water quality constituents of interest at segments 5, 32 and 86 for the timed release scenario.

The constant flow and immediate release scenarios show a decrease in the median value and the IQR range for both TDS, DOC, and TN over the base model. This indicates that predicted concentrations over the model period were lowered by the LDief water transfers. TP displayed opposite behaviour to the other constituents, which may in part be a result of the transformation equation applied to PO_4^{3-}-P in the LDief constituent file. Overall, CHLA did not change greatly from the base model due to scenario predictions showing both increases and decreases in concentrations over the simulation period, as per the time series plots.

In the timed release scenario, the LDief water was released for only 306 days (12.4% of days) with the rest of the inflows the same as the base model. This shorter duration of water transfers did not alter the overall median and IQR as markedly as the other two scenarios. Segment 86 results from the timed release scenario (Figure 7) suggest that, if the water transfers had been repeated once the 2015 freshet had subsided and continued between the subsequent freshets, then the medians and IQR range may have been much lower. Such a scenario would have been similar to the immediate release scenario but with transfers stopped temporarily each year during flood periods.

Overall, the scenario results implied that adding additional transfers of water from LDief resulted in a greater number of days with lower levels of nutrients. As we anticipated, the influx of better-quality source water from LDief improved the general water quality of BPL. The constant flow scenario appeared to be the most effective at lowering total concentrations in the plotted segments at all three transfer rates.

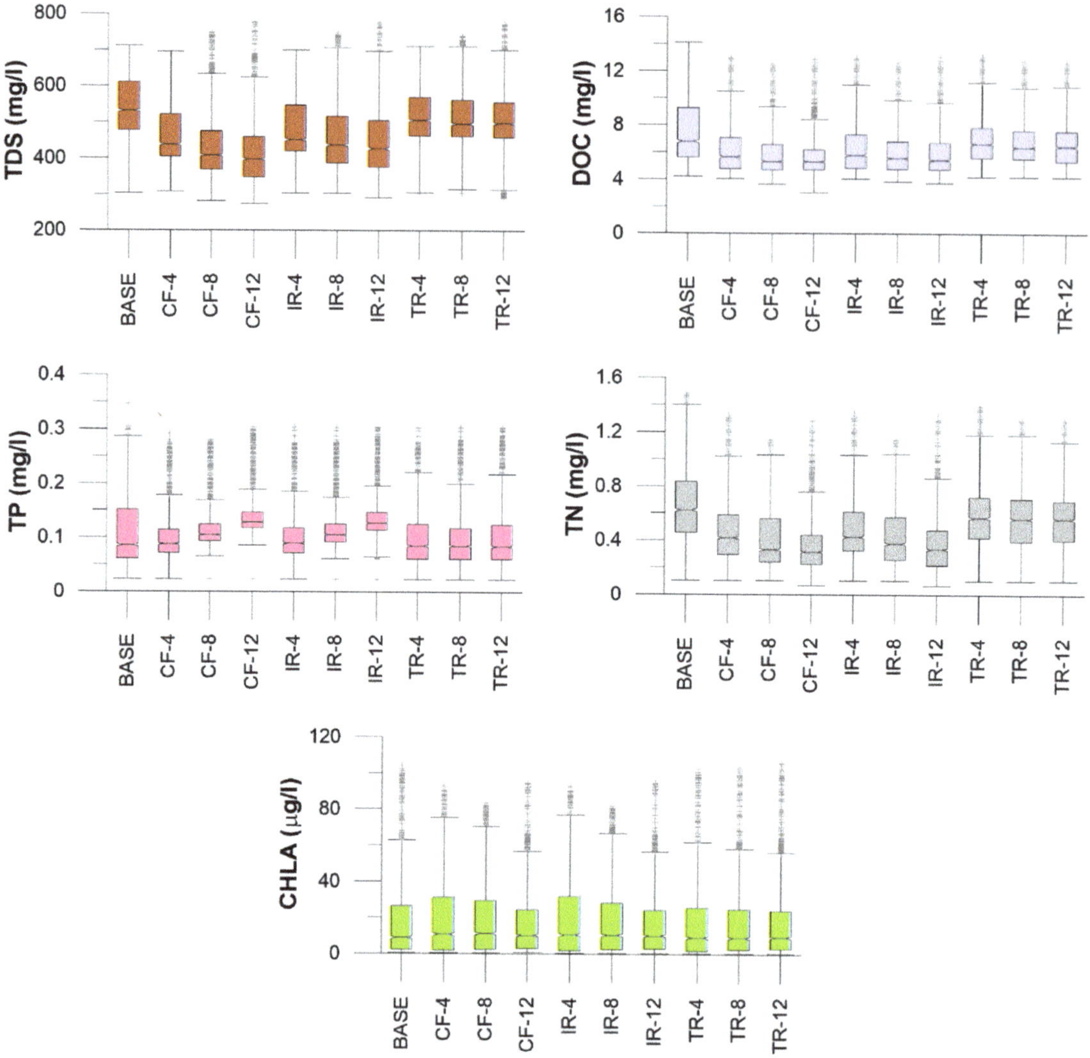

Figure 8. Boxplots comparing model output at segment 86 for the base model and the nine simulation runs for April 2013–December 2019. The box represents 50% of the ranked data (interquartile range, or IQR) with the median marked as a horizontal line within the box. Whiskers represent upper and lower bounds at (IQR $*$ factor of ±1.5).

At the same time, the increased discharge transported constituents more rapidly and further through the lake during peak events. This created shorter, more frequent peaks at the location of the BPWTP intake pipe than occurred with the base model—with the freshet nutrient peaks arriving earlier in the season. These results are consistent with the decrease in residence time and increased flow velocity from adding the additional discharge volume in the scenarios. The timed release scenario results, at segment 86 for the 2015 freshet, suggest that pausing the water transfers during periods of high flow may reduce the intensity of the nutrient peak arriving at the BPWTP intake location. It may be that trying to push as much water as one can at any time is not always effective at attaining water quality goals. In addition, pausing the water transfers during high flow events will need to occur to ensure flows remain within the Upper Qu'Appelle's channel capacity.

4. Concluding Remarks

Stakeholder engagement at the model design stage helps give confidence in the modelling process and end results. Workshops have shown that end users have a better understanding of what is being modelled, modelling possibilities and limitations, e.g., [54,55]. Stakeholders are more satisfied with output because model scenarios have been tailored to meet their own specific objectives. In this study, the provincial water management agency, the WSA, was provided with the opportunity to refine the scope of the model development and scenarios in order to address pertinent questions about water diversion strategy. For example, the WSA were keen to incorporate the extension to the model bathymetry to remove uncertainties in results brought to light in earlier studies, e.g., [32]. In turn, the enhanced in-lake water quality sampling by the WSA allowed this bathymetry to be implemented and tested in the model. The result was a more robust model that satisfies both stakeholder and scientific objectives. With this study we are bridging the needs of the end user and water quality modelling.

The modelled results show that, dependent on the timing and quantity of water transferred, the increased volume of water is predicted to decrease some water quality parameters in BPL. A flow rate of 12 m^3/s was most effective at lowering total concentrations for all scenarios. BPL's residence time is highly variable and scenario results indicate that the lake will respond rapidly to increased discharge velocity. The expansion of the model grid, and inclusion of the newer profile data, removed the uncertainties existing in the original BPL model with regards to hydrology and transport. There is strong support that the timing of LDief water transfers is important. Results for the timed release scenario suggest the optimum timing would be to wait until overland runoff is at a minimum before commencing water transfers, followed by terminating transfers before runoff resurges. Constituent concentrations near the treatment plant intake are lowest during the 2015 freshet for this scenario (Figure 7). Another benefit to this approach is that the risk of flooding the Upper Qu'Appelle system during natural high runoff events is reduced.

The difference in TP behaviour with and without the transformation equation underscores the need to understand how constituents change along the Upper Qu'Appelle before releasing any transfers. Such a study would require higher-frequency sampling data along the Upper Qu'Appelle than is available at present. Options include extending the W2 model upstream to LDief, or the coupling of two or more water quality models in a river–lake chain. The simulations in this study were based on current environmental conditions and should be treated with caution for long-term planning. Climate is changing in the Prairies and the climate change signal may eventually outweigh the benefits of releasing additional water from LDief. Testing the flow management options under climate change simulations will be the next step for the BPL model.

Author Contributions: Conceptualization, J.T., J.-M.D. and K.-E.L.; Formal analysis, J.T.; Funding acquisition, K.-E.L.; Investigation, J.T. and J.-M.D.; Methodology, J.T. and J.-M.D.; Supervision, K.-E.L.; Visualization, J.T.; Writing—original draft, J.T.; Writing—review & editing, J.-M.D. and K.-E.L. All authors have read and agreed to the published version of the manuscript.

Funding: Funding for this project was provided by the University of Saskatchewan's Global Water Futures Program.

Acknowledgments: The authors thank the Buffalo Pound Water Treatment Plant and the Saskatchewan Water Security Agency for providing water quality data used to build and verify the model. The University of Saskatchewan's School of Environment and Sustainability and Global Institute for Water Security also provided data and technical support. Analyses presented in this paper do not necessarily reflect the opinions or official position of the Saskatchewan Water Security Agency or the Government of Saskatchewan.

Conflicts of Interest: The authors declare no conflict of interest. The funders had no role in the design of the study; in the collection, analyses, or interpretation of data; in the writing of the manuscript, or in the decision to publish the results.

References

1. Castelletti, A.; Yajima, H.; Giuliani, M.; Soncini-Sessa, R.; Weber, E. Planning the Optimal Operation of a Multioutlet Water Reservoir with Water Quality and Quantity Targets. *J. Water Resour. Plan. Manag.* **2014**, *140*, 496–510. [CrossRef]
2. Zheng, T.; Sun, S.; Liu, H.; Xia, Q.; Zong, Q. Optimal control of reservoir release temperature through selective withdrawal intake at hydropower dam. *Water Sci. Technol. Water Supply* **2017**, *17*, 279–299. [CrossRef]
3. Carr, M.K.; Sadeghian, A.; Lindenschmidt, K.-E.; Rinke, K.; Morales-Marin, L. Impacts of Varying Dam Outflow Elevations on Water Temperature, Dissolved Oxygen, and Nutrient Distributions in a Large Prairie Reservoir. *Environ. Eng. Sci.* **2019**, *37*, 78–97. [CrossRef] [PubMed]
4. Weber, M.; Boehrer, B.; Rinke, K. Minimizing environmental impact whilst securing drinking water quantity and quality demands from a reservoir. *River Res. Appl.* **2019**, *35*, 365–374. [CrossRef]
5. Zhang, C.; Gao, X.; Wang, L.; Chen, Y. Analysis of agricultural pollution by flood flow impact on water quality in a reservoir using a three-dimensional water quality model. *J. Hydroinform.* **2013**, *15*, 1061–1072. [CrossRef]
6. Mamani Larico, A.J.; Zúñiga Medina, S.A. Application of WASP model for assessment of water quality for eutrophication control for a reservoir in the Peruvian Andes. *Lakes Reserv. Res. Manag.* **2019**, *24*, 37–47. [CrossRef]
7. Ziemińska-Stolarska, A.; Kempa, M. Modeling and monitoring of hydrodynamics and surface water quality in the sulejów dam reservoir, poland. *Water* **2021**, *13*, 296. [CrossRef]
8. Jeznach, L.C.; Hagemann, M.; Park, M.-H.; Tobiason, J.E. Proactive modeling of water quality impacts of extreme precipitation events in a drinking water reservoir. *J. Environ. Manag.* **2017**, *201*, 241–251. [CrossRef]
9. Morales-Marín, L.A.; Wheater, H.S.; Lindenschmidt, K.E. Assessment of nutrient loadings of a large multipurpose prairie reservoir. *J. Hydrol.* **2017**, *550*, 166–185. [CrossRef]
10. Deus, R.; Brito, D.; Mateus, M.; Kenov, I.; Fornaro, A.; Neves, R.; Alves, C.N. Impact evaluation of a pisciculture in the Tucuruí reservoir (Pará, Brazil) using a two-dimensional water quality model. *J. Hydrol.* **2013**, *487*, 1–12. [CrossRef]
11. Jeznach, L.C.; Jones, C.; Matthews, T.; Tobiason, J.E.; Ahlfeld, D.P. A framework for modeling contaminant impacts on reservoir water quality. *J. Hydrol.* **2016**, *537*, 322–333. [CrossRef]
12. Akomeah, E.; Morales-Marın, L.A.; Carr, M.; Sadeghian, A.; Lindenschmidt, K.E. The impacts of changing climate and streamflow on nutrient speciation in a large Prairie reservoir. *J. Environ. Manag.* **2021**, *288*, 112262. [CrossRef] [PubMed]
13. Shiklomanov, I.A. Appraisal and assessment of world water resources. *Water Int.* **2000**, *25*, 11–32. [CrossRef]
14. Simonovic, S.P.; Rajasekaram, V. Integrated analyses of Canada's Water resources: A system dynamics approach. *Can. Water Resour. J.* **2004**, *29*, 223–250. [CrossRef]
15. Wheater, H.; Gober, P. Water security in the Canadian Prairies: Science and management challenges. Philosophical transactions. *Ser. A Math. Phys. Eng. Sci.* **2013**, *371*, 20120409. [CrossRef]
16. Martz, L.; Bruneau, J.; Rolfe, J. *Climate Change and Water: SSRB (South Saskatchewan River Basin) Final Technical Report*; University of Saskatchewan: Saskatoon, SK, Canada, 2007.
17. Patrick, R.J. Enhancing water security in Saskatchewan, Canada: An opportunity for a water soft path. *Water Int.* **2011**, *36*, 748–763. [CrossRef]
18. Gober, P.; Wheater, H.S. Socio-hydrology and the science-policy interface: A case study of the Saskatchewan River basin. *Hydrol. Earth Syst. Sci.* **2014**, *18*, 1413–1422. [CrossRef]
19. Diamond, L. Comparative Approaches in Lake Management Planning (Chandos and Buffalo Pound Lake, Canada). *Landsc. J.* **1984**, *3*, 61–71. [CrossRef]
20. Davis, E.; Sauer, E.K. Water Quality Deterioration During Interbasin Transfer in Saskatchewan. *Can. Water Resour. J.* **1988**, *13*, 50–59. [CrossRef]
21. McGowan, S.; Leavitt, P.; Hall, R. A Whole-Lake Experiment to Determine the Effects of Winter Droughts on Shallow Lakes. *Ecosystems* **2005**, *8*, 694–708. [CrossRef]
22. Kehoe, M.; Chun, K.; Baulch, H. Who Smells? Forecasting Taste and Odor in a Drinking Water Reservoir. *Environ. Sci. Technol.* **2015**, *49*, 10984. [CrossRef] [PubMed]
23. Hosseini, N.; Akomeah, E.; Davis, J.-M.; Baulch, H.; Lindenschmidt, K.-E. Water quality modeling of a prairie river-lake system. *Environ. Sci. Pollut. Res. Int.* **2018**, *25*, 31190–31204. [CrossRef] [PubMed]
24. Cavaliere, E.; Baulch, H.M. Winter in two phases: Long-term study of a shallow reservoir in winter. *Limnol. Oceanogr.* **2021**, *66*, 1335–1352. [CrossRef]
25. Barica, J. Water Quality Problems Associated with High Productivity of Prairie Lakes in Canada: A Review. *Water Qual. Bull.* **1987**, *12*, 107–115.
26. Donald, D.B.; Parker, B.R.; Davies, J.-M.; Leavitt, P.R. Nutrient sequestration in the Lake Winnipeg watershed. *J. Great Lakes Res.* **2015**, *41*, 630–642. [CrossRef]
27. North, R.L.; Davies, J.-M.; Doig, L.E.; Lindenschmidt, K.-E.; Hudson, J.J. Lake Diefenbaker: The prairie jewel. *J. Great Lakes Res.* **2015**, *41*, 1–7. [CrossRef]

28. Acharya, M.; Kells, J.A. Conveyance capacity investigations for the Upper Qu'Appelle River Channel. In Proceedings of the 17th Canadian Hydrotechnical Conference, Edmonton, AB, Canada, 17–19 August 2015.
29. Parsons, G.F.; Thorp, T.; Kulshreshtha, S.; Gates, C. *Upper Qu'Appelle Water Supply Project: Economic Impact & Sensitivity Analysis*; Clifton Associates Ltd.: Regina, SK, Canda, 2012; p. 101.
30. Lindenschmidt, K.-E.; Davies, J.-M. Winter flow testing of the Upper Qu'Appelle River. In Proceedings of the 17th CRIPE Workshop on the Hydraulics of Ice Covered Rivers, Edmonton, AB, Canada, 21–24 July 2013; pp. 312–328.
31. Lindenschmidt, K.-E.; Carstensen, D. The upper Qu'Appelle water supply project in Saskatchewan, Canada: Upland canal ice study. *Osterr. Wasser Abfallwirtsch.* **2015**, *67*, 230–239. [CrossRef]
32. Terry, J.A.; Lindenschmidt, K.-E. Sensitivity of boundary data in a shallow prairie lake model. *Can. Water Resour. J.* **2020**, *45*, 204–215. [CrossRef]
33. Environment and Climate Change Canada. Climate Normals 1981–2010 Station Data for Moose Jaw A Station. Available online: http://climate.weather.gc.ca (accessed on 21 November 2014).
34. Water Security Agency. Dams & Reservoirs. Available online: https://www.wsask.ca/lakes-rivers/dams-reservoirs/ (accessed on 3 October 2021).
35. Sauchyn, D.J.; Barrow, E.M.; Hopkinson, R.F.; Leavitt, P.R. Aridity on the Canadian Plains. *Géographie Phys. Et Quat.* **2002**, *56*, 247–259. [CrossRef]
36. Chun, K.P.; Wheater, H.S.; Nazemi, A.; Khaliq, M.N. Precipitation downscaling in Canadian Prairie Provinces using the LARS-WG and GLM approaches. *Can. Water Resour. J.* **2013**, *38*, 311–332. [CrossRef]
37. Buffalo Pound Water Administration Board. Annual Report 2015. Available online: https://www.buffalopoundwtp.ca/images/docs/2015_buffalo_pound_annual_report.pdf (accessed on 19 July 2017).
38. Wells, S.A. *CE-QUAL-W2: A Two-Dimensional, Laterally Averaged, Hydrodynamic and Water Quality Model, Version 4.2, User Manual, Parts 1–5*; Department of Civil and Environmental Engineering, Portland State University: Portland, OR, USA, 2020.
39. Terry, J.A.; Sadeghian, A.; Lindenschmidt, K.-E. Modelling Dissolved Oxygen/Sediment Oxygen Demand under Ice in a Shallow Eutrophic Prairie Reservoir. *Water* **2017**, *9*, 131. [CrossRef]
40. Vandergucht, D.M.; Perez-Valdivia, C.; Davies, J.-M. *Qu'Appelle Nutrient Mass Balance Report 2013–2016*; Report WQ201812-01; Water Security Agency: Moose Jaw, SK, Canada, 2018.
41. Terry, J.A.; Sadeghian, A.; Baulch, H.M.; Chapra, S.C.; Lindenschmidt, K.-E. Challenges of modelling water quality in a shallow prairie lake with seasonal ice cover. *Ecol. Model.* **2018**, *384*, 43–52. [CrossRef]
42. Newbold, J.D.; Elwood, J.W.; O'Neill, R.V.; Winkle, W.V. Measuring Nutrient Spiralling in Streams. *Can. J. Fish. Aquat. Sci.* **1981**, *38*, 860–863. [CrossRef]
43. Ensign, S.H.; Doyle, M.W. Nutrient spiraling in streams and river networks. *J. Geophys. Res. Biogeosci.* **2006**, *111*, G04009. [CrossRef]
44. Lindenschmidt, K.-E. *Winter Flow Testing of the Upper Qu'Appelle River*; Lambert Academic Publishing: Saarbrucken, Germany, 2014.
45. House, W.A. Geochemical cycling of phosphorus in rivers. *Appl. Geochem.* **2003**, *18*, 739–748. [CrossRef]
46. Jarvie, H.P.; Jürgens, M.D.; Williams, R.J.; Neal, C.; Davies, J.J.L.; Barrett, C.; White, J. Role of river bed sediments as sources and sinks of phosphorus across two major eutrophic UK river basins: The Hampshire Avon and Herefordshire Wye. *J. Hydrol.* **2005**, *304*, 51–74. [CrossRef]
47. Bowes, M.J.; House, W.A.; Hodgkinson, R.A. Phosphorus dynamics along a river continuum. *Sci. Total Environ.* **2003**, *313*, 199–212. [CrossRef]
48. Butler, B.A.; Ford, R.G. Evaluating Relationships Between Total Dissolved Solids (TDS) and Total Suspended Solids (TSS) in a Mining-Influenced Watershed. *Mine Water Environ.* **2018**, *37*, 18–30. [CrossRef]
49. Rohlfs, A.-M.; Mitrovic, S.M.; Williams, S.; Hitchcock, J.N.; Rees, G.N. Dissolved organic carbon delivery from managed flow releases in a montane snowmelt river. *Aquat. Sci.* **2016**, *78*, 793–807. [CrossRef]
50. Hernes, P.J.; Spencer, R.G.M.; Dyda, R.Y.; Pellerin, B.A.; Bachand, P.A.M.; Bergamaschi, B.A. The role of hydrologic regimes on dissolved organic carbon composition in an agricultural watershed. *Geochim. Et Cosmochim. Acta* **2008**, *72*, 5266–5277. [CrossRef]
51. Fulweiler, R.W.; Nixon, S.W. Export of Nitrogen, Phosphorus, and Suspended Solids from a Southern New England Watershed to Little Narragansett Bay. *Biogeochemistry* **2005**, *76*, 567–593. [CrossRef]
52. Terry, J.A.; Mc Benskin, C.; Eastoe, E.F.; Haygarth, P.M. Temporal dynamics between cattle in-stream presence and suspended solids in a headwater catchment. *Environ. Sci. Processes Impacts* **2014**, *16*, 157–1577. [CrossRef] [PubMed]
53. Perks, M.T.; Owen, G.J.; Benskin, C.M.H.; Jonczyk, J.; Deasy, C.; Burke, S.; Reaney, S.M.; Haygarth, P.M. Dominant mechanisms for the delivery of fine sediment and phosphorus to fluvial networks draining grassland dominated headwater catchments. *Sci. Total Environ.* **2015**, *523*, 178–190. [CrossRef] [PubMed]
54. Hassanzadeh, E.; Strickert, G.; Morales-Marin, L.; Noble, B.; Baulch, H.; Shupena-Soulodre, E.; Lindenschmidt, K.-E. A framework for engaging stakeholders in water quality modeling and management: Application to the Qu'Appelle River Basin, Canada. *J. Environ. Manag.* **2019**, *231*, 1117–1126. [CrossRef]
55. Lindenschmidt, K.-E.; Akomeah, E.; Baulch, H.; Boyer, L.; Davies, J.-M.; Hassanzadeh, E.; Marin, L.M.; Strickert, G.; Wauchope, M. Interfacing Stakeholder Involvement into a Surface Water-Quality Modelling System for Water Management and Policy Development. In *New Trends in Urban Drainage Modelling*; Springer: Berlin/Heidelberg, Germany, 2019; pp. 312–316.

water

Article

Ammonium Nitrogen Streamflow Transport Modelling and Spatial Analysis in Two Chinese Basins

Jingchen Yin [1], Haitao Chen [1], Yuqiu Wang [1,*], Lifeng Guo [2,*], Guoguang Li [3] and Puzhou Wang [4]

1 College of Environmental Science and Engineering, Nankai University, Tianjin 300350, China; yinjingchen@mail.nankai.edu.cn (J.Y.); 1120210300@mail.nankai.edu.cn (H.C.)
2 Ecological Environment Monitoring and Scientific Research Center of Haihe River Basin and Beihai Sea Area, Ministry of Ecological Environment, Tianjin 300061, China
3 Shenzhen Qianming Technology Co., Ltd., Shenzhen 518000, China; liguoguang@mail.nankai.edu.cn
4 Synthego Corporation, Redwood City, CA 94063, USA; puzhou.wang@synthego.com
* Correspondence: yqwang@nankai.edu.cn (Y.W.); guolifeng@hhbhjg.mee.gov.cn (L.G.)

Abstract: Ammonium nitrogen (NH_4^+-N), which naturally arises from the decomposition of organic substances through ammonification, has a tremendous influence on local water quality. Therefore, it is vital for water quality protection to assess the amount, sources, and streamflow transport of NH_4^+-N. SPAtially Referenced Regressions on Watershed attributes (SPARROW), which is a hybrid empirical and mechanistic modeling technique based on a regression approach, can be used to conduct studies of different spatial scales on nutrient streamflow transport. In this paper, the load and delivery of NH_4^+-N in Poyang Lake Basin (PLB) and Haihe River Basin (HRB) were estimated using SPARROW. In PLB, NH_4^+-N load streamflow transport originating from point sources and farmland accounted for 41.83% and 32.84%, respectively. In HRB, NH_4^+-N load streamflow transport originating from residential land and farmland accounted for 40.16% and 36.75%, respectively. Hence, the following measures should be taken: In PLB, it is important to enhance the management of the point sources, such as municipal and industrial wastewater. In HRB, feasible measures include controlling the domestic pollution and reducing the usage of chemical fertilizers. In addition, increasing the vegetation coverage of both basins may be beneficial to their nutrient management. The SPARROW models built for PLB and HRB can serve as references for future uses for different basins with various conditions, extending this model's scope and adaptability.

Keywords: Poyang Lake Basin; Haihe River Basin; SPARROW; ammonium nitrogen; nutrient transport

Citation: Yin, J.; Chen, H.; Wang, Y.; Guo, L.; Li, G.; Wang, P. Ammonium Nitrogen Streamflow Transport Modelling and Spatial Analysis in Two Chinese Basins. *Water* **2022**, *14*, 209. https://doi.org/10.3390/w14020209

Academic Editor: Karl-Erich Lindenschmidt

Received: 20 November 2021
Accepted: 7 January 2022
Published: 11 January 2022

1. Introduction

An excess of nutrients, which are generated from point and non-point sources and are eventually transported to water bodies, has led to severe eutrophication throughout the world in recent years [1]. Eutrophication of water bodies causes toxic algal blooms, oxygen depletion, loss of biodiversity, and thereby, the degradation of water quality and aquatic ecosystem services [2]. Ammonium nitrogen (NH_4^+-N), which naturally arises from the decomposition of organic substances through ammonification, is a critical nutrient produced by human activities such as fertilizing, livestock breeding, and municipal wastewater treating. Synthetic nitrogen fertilizers are slightly absorbed by crops (about 10%), while large quantities of synthetic nitrogen are exported to aquatic systems through surface runoff and decomposed to NH_4^+-N [3]. Domestic and industrial wastewater are also important manners, by which NH_4^+-N enters into aquatic systems. Excessive NH_4^+-N leads to eutrophication, endangering aquatic species and polluting water sources [4]. Since NH_4^+-N has a tremendous influence on local water quality, it is vital for water quality protection to assess the amount, sources, and transport of NH_4^+-N [5].

Previous research has focused on the processes of NH_4^+-N stream transport. For instance, Jin et al. employed the Integrated Nitrogen Catchments (INCA-N) model to link upstream processes to downstream water quality of NH_4^+-N in the Hampshire Avon catchment [6]. Ervinia et al. applied the INCA-N model to identify the source and processes of NH_4^+-N in the Jiulong River Watershed (JRW) [7]. Zhang et al. employed QUAL2K model to explore the transport of NH_4^+-N in a creek watershed with sparse data in southeast of China [8]. Xue et al. simulated the land surface hydrological runoff and routing processes in the Xiaoqing River Basin and analysed the concentration of NH_4^+-N temporally and spatially, using the Soil and Water Assessment Tool (SWAT) model and HEC-RAS model [9]. Dai et al. assessed the sources and transport of ammonium nitrogen in a karst basin using the SPAtially Referenced Regressions on Watershed attributes (SPARROW) model [5].

Complex models, such as Agricultural Non-point Source (AGNPS), Hydrological Simulation Program FORTRAN (HSPF), INCA-N, and SWAT, which have been developed to evaluate water quality and sources of nutrients [10], have hefty data requirements [11] and are time-consuming processes [12]. Hybrid empirical and mechanistic models based on regression, such as the SPARROW model, can be used to conduct studies at different spatial scales on nutrient transport with smaller input datasets, such as data on nutrient load, nutrient sources, and landscape properties. Since it was first established by the U.S. Geological Survey (USGS) [13], SPARROW has been extensively applied in North America [14–24], Asia [25–30], New Zealand [31,32], Spain [33,34], and Brazil [35] with satisfactory performance. SPARROW has been employed in studies on regions of various sizes, from 153 km^2 [36] to 3.2 million km^2 [37]. Furthermore, the model performs well in both estimating the influences of human activities on the environment [38,39], and analyzing scenarios of climate change [40,41] or land use change [42,43].

Poyang Lake Basin (PLB) consists of five main river watersheds. Its water exchanges with Yangtze River after the streamflow of the five main rivers are injected into Poyang Lake, the largest freshwater lake in China [44]. PLB is a typical southern water basin in China, since it has a massive quantity of water and better water quality than northern water basins in China. Haihe River Basin (HRB), which contains seven major river watersheds, has a high population density and numerous large cities. Therefore, it plays an important role in the politics and economy of China [45]. The water quantity and quality of HRB is seriously affected by human activities, such as irrigation, fertilization, point sources, etc. The main purposes of this paper were to (1) establish SPARROW models in PLB and HRB and figure out the NH_4^+-N load and streamflow transport in these two basins; (2) compare NH_4^+-N load and streamflow transport between PLB and HRB and identify the distinctions between the two basins; (3) offer assistance in future control measures for better management in PLB and HRB.

2. Materials and Methods

2.1. Study Area

PLB is located in southern China, between 113° E–118° E and 24° N–30° N, as shown in Figure 1a. The drainage area of PLB is 162,271 km^2, of which 156,743 km^2 is located in Jiangxi Province (the total land area of Jiangxi Province is 166,946 km^2), accounting for 97% of the drainage area of PLB and 94% of the total land area of Jiangxi Province [46]. PLB belongs to a subtropical warm and humid monsoon climate zone. The average annual temperature is about 16.3–19.5 °C, generally increasing from north to south. The vegetation type is subtropical, evergreen, broad-leaved forest. PLB is composed of five main tributary rivers, namely, Ganjiang River, Fuhe River, Xinjiang River, Raohe River, and Xiushui River, as well as several smaller rivers. These rivers finally flow into Poyang Lake, and the outflow of Poyang Lake inject into Yangtze River through the Hukou station. The span of PLB covers a length of 620 km from north to south and a width of 490 km from east to west. The basin is surrounded by mountains in the east, west, and south; has hills and valley plains crisscrossing the central part; and there is the Poyang Lake plain in the north. The terrain of PLB is high in the south and low in the north, which is conducive to water convergence.

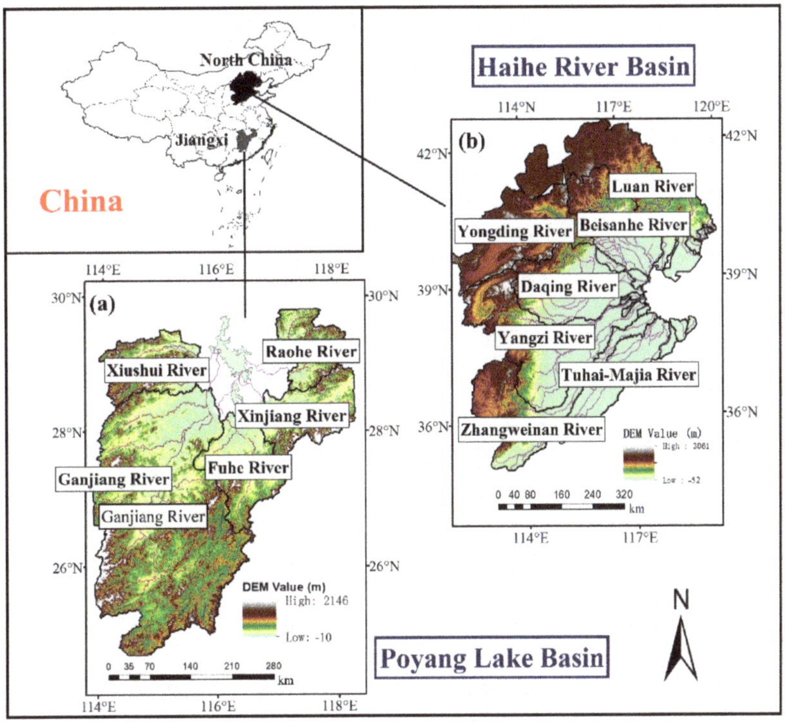

Figure 1. Location and Digital Elevation Model (DEM) of (**a**) PLB and (**b**) HRB.

The annual average total amount of water resources in PLB was 163.06 billion m^3, while the annual average total amount of water uses is 17.67 billion m^3 and the annual average quantity of wastewater is 4.39 billion m^3. The total water resources of Ganjiang River, Fuhe River, Xinjiang River, Raohe River, and Xiushui River are 89.13, 20.19, 21.11, 15.77, and 16.86 billion m^3, respectively. The average annual runoff in PLB into Poyang Lake is 142.73 billion m^3. The annual average runoff totals of Ganjiang River, Fuhe River, Xinjiang River, Raohe River, and Xiushui River into Poyang Lake are 81.26, 16.20, 18.53, 14.04, and 12.70 billion m^3, respectively.

HRB is located in northern China, between 112° E–120° E and 35° N–43° N, as shown in Figure 1b. The scope of HRB covers eight provinces (or cities). The total area of Beijing and Tianjin, 91% of the area of Hebei province, 38% of the area of Shanxi Province, 20% of the area of Shandong Province, 9.2% of the area of Henan province, 13,600 km^2 of Inner Mongolia Autonomous Region, and 1700 km^2 of Liaoning Province belong to Haihe River Basin—318,200 km^2 in total. The climate of HRB is semi-humid and semi-arid. It is located in the temperate East Asian monsoon climate area. The annual average temperature is 1.5–14.0 °C [47]. The vegetation types in HRB are various due to the monsoon climate. HRB is constituted by seven major river watersheds, Luan River, Beisanhe River, Yongding River, Daqing River, Ziya River, Zhangweinan Canal, and Tuhai-Majia River. These rivers mostly flow from west to east, and finally discharge into Bohai Bay. HRB is bordered by the Shanxi Plateau and Yellow River Basin in the west, the Mongolia Plateau and the inland river basin in the north, the Yellow River in the south, and the Bohai Sea in the east. The total terrain of HRB is high in the northwest and low in the southeast, which is composed of three landforms: plateau, mountain, and plain. Plateaus and mountains are located in the north and west of HRB, covering an area of 189,400 km^2, accounting for 60% of the basin area. The east and southeast of HRB are covered by a plain, which covers

128,400 km^2, accounting for 40% of the basin's area. The annual average total amount of water resources in HRB was 30.30 billion m^3, while the annual average total amount of water uses is 37.00 billion m^3 and the annual average quantity of wastewater is 5.98 billion m^3. HRB belongs to an area with an extreme water shortage. The total amount of inter basin water transfer in HRB was 3.51 billion m^3 (including the amount of water diverted from Yangtze River and Yellow River). The total water resources of Luanhe River, Beisanhe River, Yongding River, Daqing River, Ziya River, Zhangweinan Canal, and Tuhai-Majia River are 4.67, 4.38, 2.49, 4.65, 6.36, 4.41, and 3.32 billion m^3, respectively. The average annual runoff in HRB into Bohai Bay was 3.51 billion m^3. The annual average runoff totals of Luanhe River, Beisanhe River, Daqing River, Ziya River, and Tuhai-Majia River into Bohai Bay are 0.92, 1.05, 0.64, 0.34, and 0.54 billion m^3, respectively. It is obvious that PLB has far more water resources than HRB, especially the amounts of water delivered to the outlets. Such difference might be explained by the lesser amounts of precipitation, the larger area of plain terrain, and the larger demand for water supply in HRB, leading to the lesser NH_4^+-N loads delivered to the outlets compared to PLB. The summary of water resources of the main rivers in PLB and HRB is in Tables S1 and S2.

2.1.1. Nutrient Sources

Point source data in 2017 of NH_4^+-N in PLB and HRB were provided by Chinese Research Academy of Environmental Sciences and Ecological Environment Monitoring and Scientific Research Center of Haihe River Basin and Beihai Sea Area, respectively. The load from point sources involved municipal and industrial wastewater treatment plants. The median annual load in PLB from point sources was 125,744 kg/year of NH_4^+-N, with a range of 4479–1,568,808 kg/year. The median annual load in HRB from point sources was 42,084 kg/year of NH_4^+-N, with a range of 0–2,952,298 kg/year.

Farmland, woodland, grassland, and residential land were considered as non-point sources of nutrients [48]. Land use data in 2015 were acquired from the Data Center for Resources and Environmental Sciences of Chinese Academy of Sciences (https://www.resdc.cn/data.aspx?DATAID=184, accessed on 22 December 2020) and used to label farmland, woodland, grassland, water body, residential land, and barren land. In PLB, there were 24.60% farmland, 67.15% woodland, 4.20% grassland, 1.98% water body, 2.04% residential land, and 0.01% barren land, as shown in Figure 2a. In HRB, there were 49.03% farmland, 19.16% woodland, 18.45% grassland, 2.44% water body, 9.74% residential land, and 1.18% barren land, as shown in Figure 2b. It is clear that the areas of woodland and grassland in PLB were almost equal to those in HRB; and that the areas of farmland and residential land in PLB were far smaller than those in HRB.

2.1.2. River Network and Nutrient Load Estimates

The river network was obtained from the Digital Elevation Model (DEM) data (90 m × 90 m) of PLB and HRB by using ArcHydro tools; 85 and 310 stream reaches were delineated, respectively. The DEM data were downloaded from the Geospatial Data Cloud site, Computer Network Information Center, Chinese Academy of Sciences (http://www.gscloud.cn/sources/accessdata/306?pid=302, accessed on 28 December 2020).

Streamflow data in 2017 were available from only Qiujing, Meigang, Wanjiabu, Hushan, Waizhou, Dufengkeng, and Lijiadu stations in PLB, which were provided by the Jiangxi Academy of Environmental Sciences. Streamflow at ungauged watersheds in PLB was estimated by using a GWLF model [49]. The version of the GWLF model used in this study was ReNuMa version 2.2.2 [50]. Streamflow data were available from 203 stations in HRB, which were provided by Ecological Environment Monitoring and Scientific Research Center of Haihe River Basin and Beihai Sea Area. Streamflow at ungauged watersheds in HRB was simulated by interpolation analysis using the GIS platform. Water quantity station distributions are shown in Figure 3a,c.

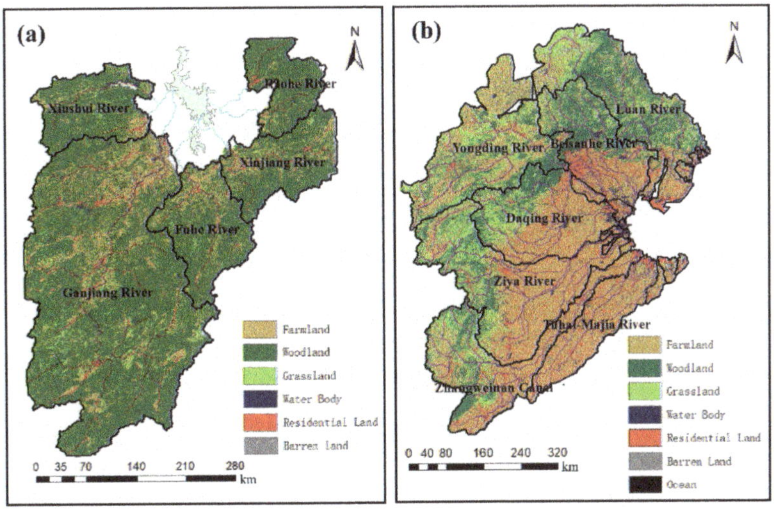

Figure 2. Land use distributions in (**a**) PLB and (**b**) HRB.

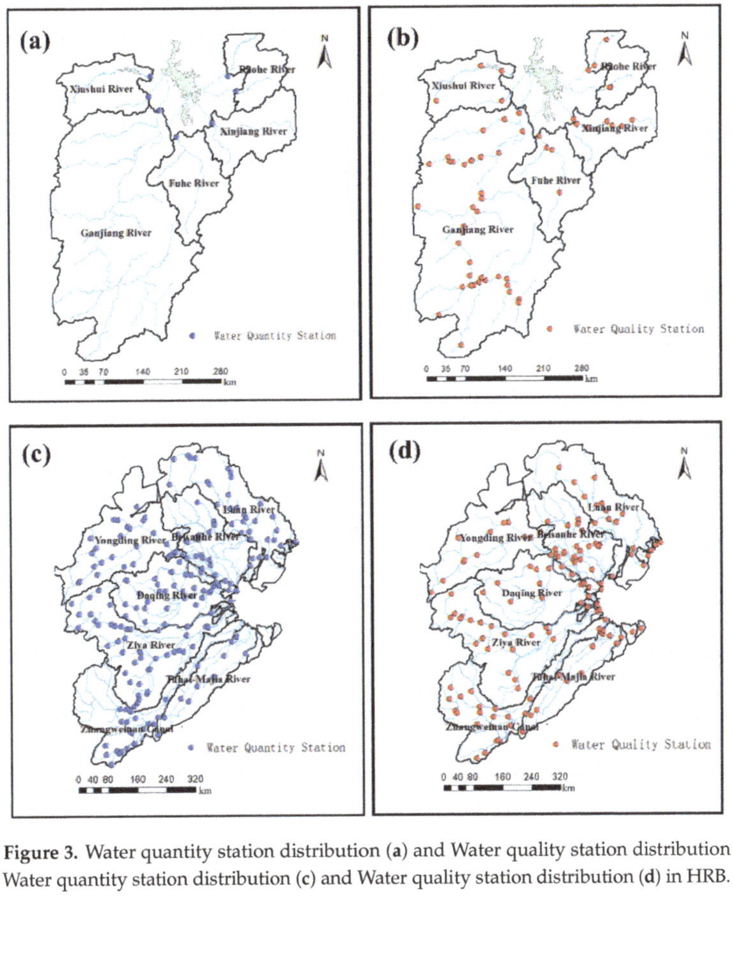

Figure 3. Water quantity station distribution (**a**) and Water quality station distribution (**b**) in PLB, Water quantity station distribution (**c**) and Water quality station distribution (**d**) in HRB.

Water quality data in 2017 were provided monthly by the Jiangxi Academy of Environmental Sciences from 58 sites for PLB, and by Ecological Environment Monitoring and Scientific Research Center of Haihe River Basin and Beihai Sea Area from 144 sites for HRB. Water quality station distributions are shown in Figure 3b,d.

Annual nutrient loads of these sites were assessed as follows:

$$Load_i = Flow_i \times Conc_i \tag{1}$$

where $Load_i$ is the Annual nutrient load of reach i, $Flow_i$ is simulated by the GWLF model (or interpolation analysis) as above, and $Conc_i$ represents the mean annual instream $NH_4{}^+$-N concentration.

2.1.3. Land-to-Water Delivery Variables

The model's land-to-water delivery variables were mainly determined by the spatial attribute data. Annual average precipitation, annual average temperature, and slope were used in model calibration. The data of precipitation and temperature were downloaded from China Meteorological Data Service Centre (http://data.cma.cn/data/cdcdetail/dataCode/SURF_CLI_CHN_MUL_DAY_V3.0.html, accessed on 19 June 2020). The slope data were extracted from DEM. The data of delivery variables were transformed to the final type of SPARROW model by using GIS platform.

2.2. SPARROW Model

The total load leaving a given reach in the SPARROW model is considered as the sum of the load produced by itself and the load delivered from its upper stream [51]. The total load leaving reach i can be expressed mathematically as follows:

$$F_i^* = \left(\sum_{j\epsilon J(i)} F_j'\right)\delta_i A\left(Z_i^S;\theta_S\right) + \left(\sum_{n=1}^{N_S} S_{n,i}\alpha_n D_n\left(Z_i^D;\theta_D\right)\right)A'\left(Z_i^S;\theta_S\right) \tag{2}$$

where F_i* is the total load leaving reach i (kg/year), F_j' is the load leaving upstream reaches from reach j, the set $J(i)$ is all upstream reaches of reach i, δ_i is the proportion of load delivered to reach i contributed by adjacent upstream reaches, $A(Z_i^S;\theta_S)$ is a function of first-order loss processes related to stream size, and $A'(Z_i^S;\theta_S)$ is the square root of $A(Z_i^S;\theta_S)$.

$S_{n,j}$ is the nutrient source n in reach i, N_S is the total number of nutrient sources, α_n is the coefficient of nutrient source n, and the land-to-water delivery term $D_n(Z_i^D;\theta_D)$ is defined as follows:

$$D_n\left(Z_i^D;\theta_D\right) = \exp\left(\sum_{m=1}^{M_D} \omega_{nm} Z_{m\ i}^D \theta_{Dm}\right) \tag{3}$$

where $Z_{m\ i}^D$ is the land-to-water variable m within reach i, M_D is the total number of delivery variables, θ_{Dm} is the coefficient of delivery variable m, and ω_{nm} is the delivery index for judging whether source n uses delivery variable m or not.

Stream delivery function, considered as an attenuation process acting on flux, is formulated by a first-order reaction rate process. The proportion of the load remaining after the delivery to the outlet of reach i is expressed as an exponential function:

$$A\left(Z_i^S;\theta_S\right) = \exp\left(-\sum_{c=1}^{C_S} \theta_{Sc} T_{c\ i}^S\right) \tag{4}$$

where $T_{c\ i}^S$ is the average travelling time of a stream in reach i, which is classified as in-stream decay class c. c is the number of in-stream decay class c streams, C_S is the total number of in-stream decay classes, and θ_{Sc} is the coefficient corresponding to average travelling time of stream. Two in-stream decay classes were used in the PLB and HRB SPARROW models.

The estimation method of the SPARROW model is a nonlinear weighted least squares (NWLS) algorithm performed by the SAS procedure PROC MODEL, based on Equation (2). NWLS, which is a robust technique to solve nonlinear problems, can be considered as an iterative linear estimation process, since it is related to the ordinary least squares [51]. Bootstrap analysis is used to validate the model and perform the uncertainty analysis. The bootstrap procedure is executed by randomly selecting with replacement monitored loads from the observations in the original calibration data set and fitting separate regression models to the resampled data [32].

2.3. GWLF Model

GWLF is a combined distributed/lumped parameter model [52], which simulates monthly and annual streamflow, sediment transport, and associated nitrogen and phosphorus fluxes. GWLF was designed to be used in mixed-use watersheds (such as urban, multiple agricultural land uses, and forested land use). Streamflow from each land use category is parameterized using a variation of the Soil Conservation Service (SCS) curve number formulation; erosion is generated using the Universal Soil Loss Equation (USLE). The wastewaters from residences are considered a septic system component in GWLF model. The Excel Solver is used to make calibration in GWLF [53].

3. Results

3.1. Calibration and Validation of GWLF Model

The period from 2015 to 2016 was chosen for model calibration, and 2017 was used for model validation of streamflow. The parameters used during calibration were recession coefficient, seepage coefficient, and SCS curve numbers of different land use types. The optimized parameters are listed in Table S3. The Nash–Sutcliffe efficiency (NSE) is provided to analysis the model performance. Figure 4 shows the results of calibration and validation at seven gauged stations in PLB. The NSE of calibration ranges from 0.67 to 0.96, while the NSE of validation ranges from 0.51 to 0.88. Such results indicate an acceptable predictive capability of GWLF to estimate streamflow.

3.2. Calibration of SPARROW Models and Uncertainty Analysis

3.2.1. Calibration of SPARROW Models

NWLS calibration of SPARROW models in 2017 was executed, based on 58 and 144 water quality stations in PLB and HRB, respectively. The correlation coefficient (R^2), the mean square error (MSE), the root mean square error (RMSE), and the NSE (see Table 2) were assessed to evaluate the performances of the models.

Table 1. Results of SPARROW models.

Basin	Number of Observations	R^2	MSE	RMSE	NSE
PLB	58	0.89	0.25	0.50	0.88
HRB	144	0.53	1.91	1.38	0.52

For the results of SPARROW models for PLB and HRB, the values of R^2 were 0.89 and 0.53, respectively, which verified the acceptability of SPARROW models. MSE were 0.25 and 1.91, respectively. The RMSE for the models were around 0.50 and 1.38, respectively. Although the RMSE of HRB was significantly higher, similar values have been reported by other studies (RMSE = 1.40 [54], RMSE = 0.96 [16]). In addition, the NSE values were similar to the R^2 values, which indicated the robustness of the models. Figures 5a and 6a are scatterplots of predicted values versus observed values, in which the majority of the points are located in the vicinity of bisection. The scatterplots of residuals illustrated homoscedasticity (Figures 5b and 6b). The aforementioned analysis implied rational performance of the SPARROW models.

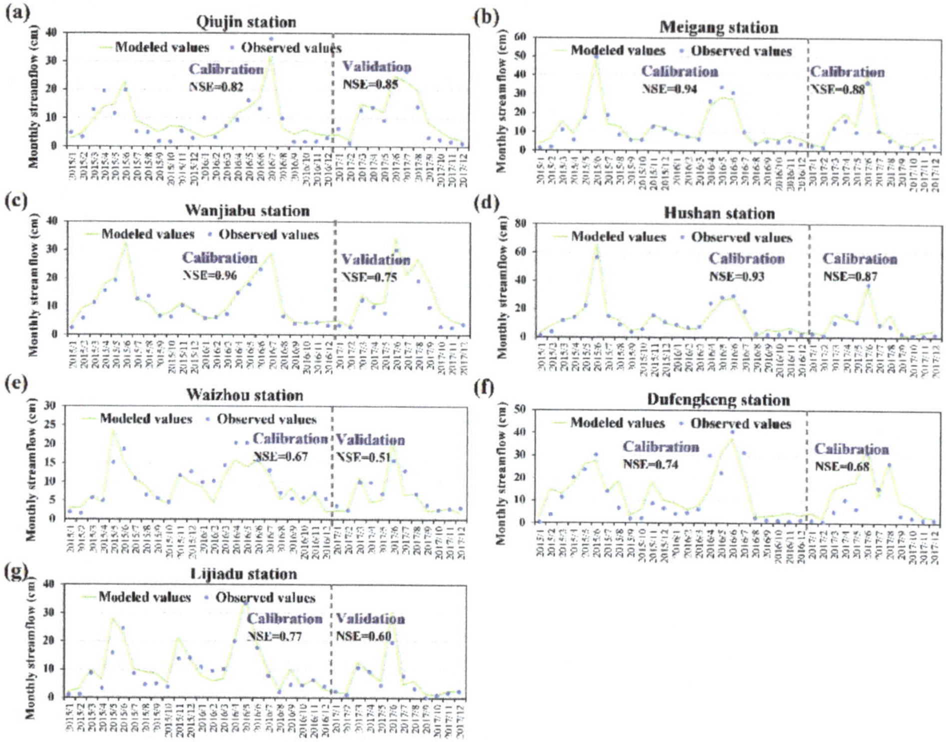

Figure 4. Modeled versus Observed values of streamflow and NSE at (**a**) Qiujing, (**b**) Meigang, (**c**) Wanjiabu, (**d**) Hushan, (**e**) Waizhou, (**f**) Dufengkeng, and (**g**) Lijiadu.

3.3. Calibration of SPARROW Models and Uncertainty Analysis

3.3.1. Calibration of SPARROW Models

NWLS calibration of SPARROW models in 2017 was executed, based on 58 and 144 water quality stations in PLB and HRB, respectively. The correlation coefficient (R^2), the mean square error (MSE), the root mean square error (RMSE), and the NSE (see Table 2) were assessed to evaluate the performances of the models.

Table 2. Results of SPARROW models.

Basin	Number of Observations	R^2	MSE	RMSE	NSE
PLB	58	0.89	0.25	0.50	0.88
HRB	144	0.53	1.91	1.38	0.52

For the results of SPARROW models for PLB and HRB, the values of R^2 were 0.89 and 0.53, respectively, which verified the acceptability of SPARROW models. MSE were 0.25 and 1.91, respectively. The RMSE for the models were around 0.50 and 1.38, respectively. Although the RMSE of HRB was significantly higher, similar values have been reported by other studies (RMSE = 1.40 [54], RMSE = 0.96 [16]). In addition, the NSE values were similar to the R^2 values, which indicated the robustness of the models. Figures 5a and 6a are scatterplots of predicted values versus observed values, in which the majority of the points are located in the vicinity of bisection. The scatterplots of residuals illustrated homoscedasticity (Figures 5b and 6b). The aforementioned analysis implied rational performance of the SPARROW models.

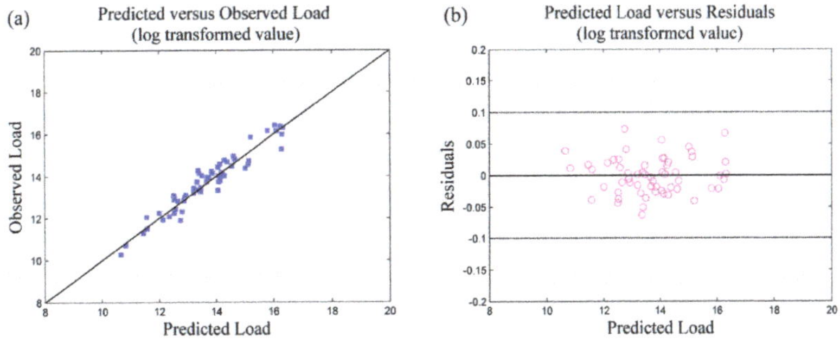

Figure 5. Scatterplots of (**a**) predicted values versus observed values and (**b**) residuals in PLB.

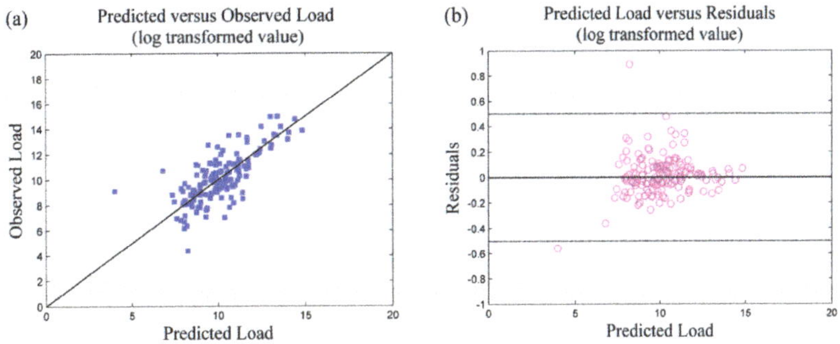

Figure 6. Scatterplots of (**a**) predicted values versus observed values and (**b**) residuals in HRB.

3.3.2. Parameters and Uncertainty Analysis

The bootstrap method was used to perform uncertainty analysis of SPARROW model parameters. Point sources, farmland, woodland and grassland, and residential land were considered as the main NH_4^+-N sources. Tables 3 and 4 show the evaluation of PLB and HRB, respectively. The coefficients of these sources all fell into the 90% confidence interval. The coefficients of farmland, woodland and grassland, and residential land were larger than 1, similar to the coefficients reported in other studies [55,56]. The coefficients of point sources were lower than 1, which suggested overestimates in the data of point sources.

Slope, average precipitation, and temperature were chosen as the land-to-water delivery variables for the SPARROW models. The coefficients of these sources all fell into the 90% confidence interval. Point sources and average precipitation showed statistical significance ($p < 0.05$) for PLB, while only average temperature showed statistical significance ($p < 0.05$) for HRB. Both stream decay variables lay in the 90% confidence intervals for PLB and HRB. Meanwhile, only the reach decay factor of a small river showed statistical significance ($p < 0.05$) for HRB.

Table 3. Evaluation of parameters in PLB.

Model Parameter	Unit	Value	Standard Error	Unbiased Value	p-Value	Lower 90% CI	Upper 90% CI
Point sources	$kg{\cdot}year^{-1}$	0.532	0.196	0.518	0.009	0.204	0.878
Farmland	$kg{\cdot}km^{-2}{\cdot}year^{-1}$	218.034	214.467	281.267	0.314	−27.661	436.068
Woodland and grassland	$kg{\cdot}km^{-2}{\cdot}year^{-1}$	24.413	41.268	31.735	0.557	−3.313	48.826
Residential land	$kg{\cdot}km^{-2}{\cdot}year^{-1}$	1101.034	2126.541	−580.387	0.607	−5085.246	2202.069
Precipitation	cm	0.003	0.001	0.003	0.002	0.002	0.004
Slope	%	−0.060	0.100	−0.087	0.551	−0.215	0.007
Small river	day^{-1}	0.070	0.133	0.052	0.599	−0.172	0.252
Large river	day^{-1}	0.017	0.104	0.008	0.873	−0.207	0.139

Table 4. Evaluation of parameters in HRB.

Model Parameter	Unit	Value	Standard Error	Unbiased Value	p-Value	Lower 90% CI	Upper 90% CI
Point sources	$kg{\cdot}year^{-1}$	0.040	0.023	0.032	0.078	0.002	0.052
Farmland	$kg{\cdot}km^{-2}{\cdot}year^{-1}$	26.838	15.649	23.760	0.089	2.564	45.117
Woodland and grassland	$kg{\cdot}km^{-2}{\cdot}year^{-1}$	6.556	5.102	5.336	0.201	−6.636	13.112
Residential land	$kg{\cdot}km^{-2}{\cdot}year^{-1}$	114.065	61.755	116.572	0.067	−9.354	226.851
Precipitation	cm	0.013	0.021	0.012	0.539	−0.014	0.034
Temperature	°C	0.148	0.072	0.132	0.042	−0.005	0.263
Small river	day^{-1}	0.438	0.084	0.423	<0.05	0.300	0.635
Large river	day^{-1}	0.114	0.081	0.094	0.165	−0.045	0.186

3.4. *Nutrient Estimates*

3.4.1. Nutrient Load

Figure 7a,b show the incremental loads of NH_4^+-N in each catchment in PLB and HRB. The incremental load ranged from 4513 to 1,468,867 kg/year in PLB. The middle reach of Ganjiang River delivered the maximum incremental NH_4^+-N load to Poyang Lake (Figure 7a). Fuhe River's downstream catchment and the upstream reach of Raohe River had large load deliveries. The upstream reach of Xinjiang River and the downstream reach of Ganjiang River had large incremental load.

In HRB, the incremental load ranged from 10 to 1,121,252 kg/year (Figure 7b). The upstream reach of Daqing River had the maximum incremental NH_4^+-N load delivery. The upstream reach of Ziya River and the middle reach of Daqing River made large incremental NH_4^+-N load deliveries. The upstream reaches of Majia River, Beisanhe River, and Zhangweinan Canal; and the downstream portions of Beisanhe River and Tuhai River, had large incremental NH_4^+-N load.

3.4.2. Nutrient Sources

Nutrient source apportionment was analyzed for the major river watersheds in PLB and HRB (Tables 5 and 6, respectively) [57]. On average, NH_4^+-N from point sources (industrial and sewage discharge), farmland, woodland and grassland, and residential land accounted for 43.09%, 32.14%, 10.70%, and 14.07%, respectively, in PLB. Point sources were the main NH_4^+-N sources in the Ganjiang, Fuhe, Xinjiang, and Xiushui River watersheds. In the Raohe River watershed, farmland was the dominant NH_4^+-N source.

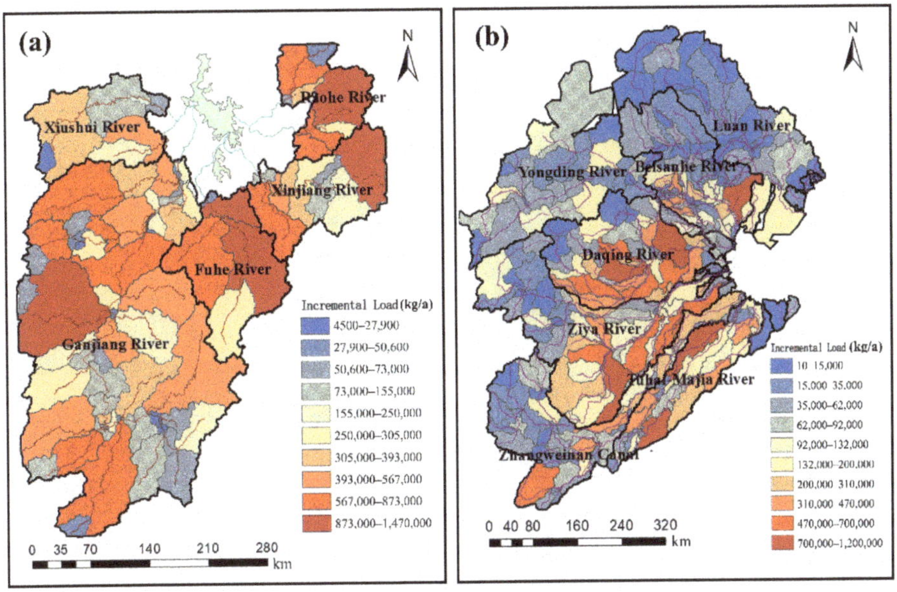

Figure 7. Incremental loads (kg/a) of NH_4^+-N in (**a**) PLB and (**b**) HRB.

Table 5. Nutrient sources' fluxes and fractions in main rivers in PLB.

River	Point Sources		Farmland		Woodland and Grassland		Residential Land		Total	
	Flux (ton/a)	Fraction (%)	Flux (ton/a)	Fraction (%)	Flux (ton/a)	Fraction (%)	Flux (ton/a)	Fraction (%)	Flux (ton/a)	Fraction (%)
Ganjiang	7005	49.48	4193	29.61	1150	8.12	1811	12.79	14158	100.00
Fuhe	1478	44.87	1146	34.81	286	8.69	383	11.63	3293	100.00
Xinjiang	1593	48.33	972	29.48	259	7.86	473	14.33	3297	100.00
Raohe	656	15.18	1764	40.85	978	22.66	921	21.32	4318	100.00
Xiushui	628	48.56	397	30.67	146	11.32	122	9.46	1293	100.00
Total	11,359	43.09	8471	32.14	2820	10.70	3709	14.07	26,360	100.00

Table 6. Nutrient sources' fluxes and fractions of main rivers in HRB.

River	Point Sources		Farmland		Woodland and Grassland		Residential Land		Total	
	Flux (ton/a)	Fraction (%)	Flux (ton/a)	Fraction (%)	Flux (ton/a)	Fraction (%)	Flux (ton/a)	Fraction (%)	Flux (ton/a)	Fraction (%)
Luan	132	14.35	299	32.48	202	21.90	288	31.28	921	100.00
Beisanhe	1449	26.71	1421	26.20	241	4.44	2313	42.65	5424	100.00
Yongding	394	19.84	846	42.61	222	11.19	523	26.35	1985	100.00
Daqing	3260	38.44	2740	32.30	197	2.32	2284	26.93	8481	100.00
Ziya	926	14.10	3253	49.52	208	3.17	2182	33.21	6569	100.00
Zhangweinan Canal	527	12.48	1909	45.21	262	6.20	1525	36.11	4223	100.00
Tuhai-Majia	913	19.29	1976	41.74	7	0.15	1837	38.81	4734	100.00
Total	7602	23.51	12,444	38.48	1339	4.14	10,952	33.87	32,336	100.00

In HRB, NH_4^+-N from point sources (industrial and sewage discharge), farmland, woodland and grassland, and residential land accounted for, on average, 23.51%, 38.48%,

4.14%, and 33.87%, respectively. In Luanhe, Yongding, Ziya, Zhangweinan Canal, and Tuhai-Majia River watersheds, farmland was the primary NH_4^+-N pollution source. In the Beisihe River watershed, residential land was the dominant NH_4^+-N source. Point sources were the main NH_4^+-N sources in the Daqing River watershed.

4. Discussion

4.1. Delivery Fraction

Figure 8 illustrates the fraction of the load that each reach delivered to the target reach through the drainage lines of rivers. The fraction of NH_4^+-N entering Poyang Lake from rivers ranged from 0.61 to 1.00, with an average of 0.86 (Figure 8a). Meanwhile, the fraction of NH_4^+-N entering Bohai Bay from rivers in HRB ranged from 0 to 0.99, with an average of 0.22 (Figure 8b). As shown in Figure 8, the delivery fraction of a specific reach is related to its distance to the target reach. In both basins, the delivery fraction decreases as the distance increases. The average delivery fraction of HRB is much lower than that of PLB, because the streamflow of PLB is much larger than that of HRB and the terrain of PLB is steeper than that of HRB.

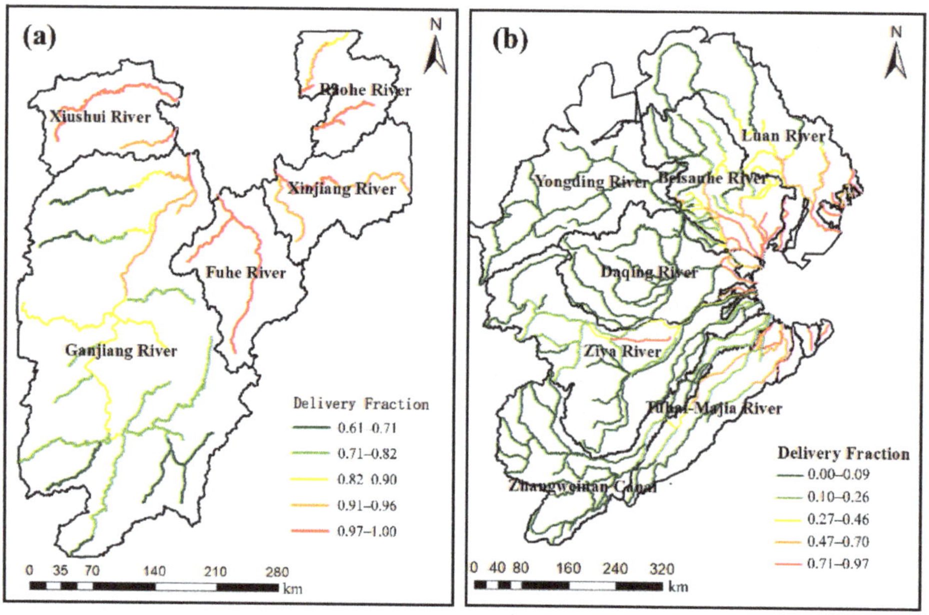

Figure 8. Delivery fractions in (**a**) PLB and (**b**) HRB.

4.2. Nutrient Delivery to Outlets

Figure 9 shows the loads and yields of NH_4^+-N delivered to the target outlets in each catchment in PLB. Figure 10 shows the loads and yields of NH_4^+-N delivered to the target outlets in each catchment in HRB. Fuhe River's downstream catchment had the maximum incremental NH_4^+-N load delivered to Poyang Lake (Figure 9a). The middle reach of Ganjiang River and the upstream reach of Raohe River had large load deliveries. The upstream reach of Xinjiang River and the downstream reach of Xiushui River had large load deliveries.

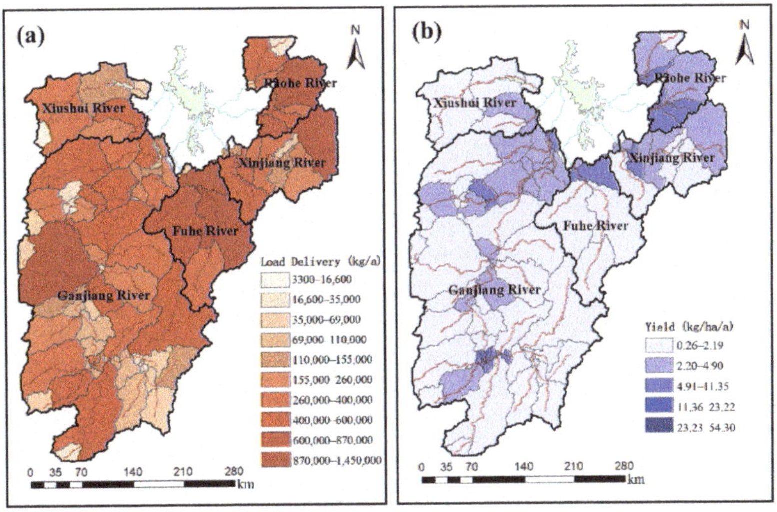

Figure 9. (**a**) Loads and (**b**) yields of NH_4^+-N delivered to the target outlets in PLB.

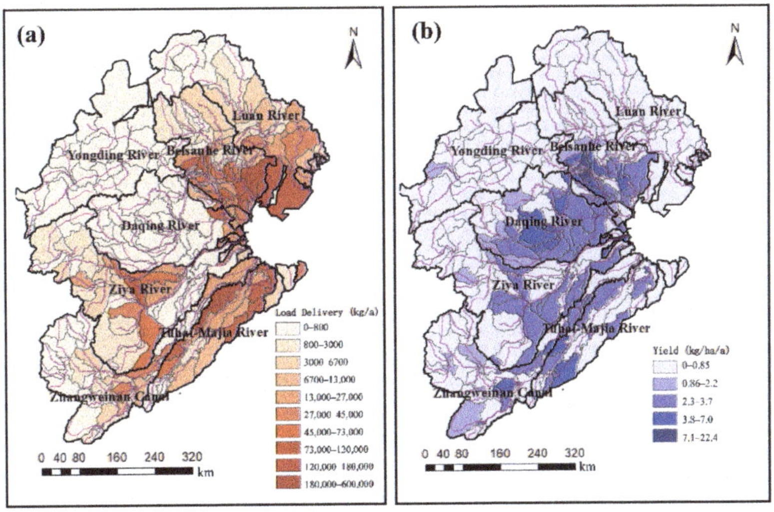

Figure 10. (**a**) Loads and (**b**) yields of NH_4^+-N delivered to the target outlets in HRB.

The yield was calculated as the division of the incremental load of a specific reach delivered to the river watershed outlet by the incremental reach area. Yield reflects the intensity of NH_4^+-N transferred to the river watershed outlets. Such analysis helps to identify major contributing areas to the water quality of outlets. As shown in Figure 9b, the middle reach of Ganjiang River had the highest NH_4^+-N yields. The downstream portions of Ganjiang River, Fuhe River, Xinjiang River, and Raohe River had high incremental NH_4^+-N loads, but the downstream portion of Ganjiang River had a larger NH_4^+-N yield than the others. These regions belong to the central urban areas of major cities in Jiangxi Province, which is perhaps the reason why these catchments have high yields.

In HRB, the downstream reach of Beisanhe River had the maximum incremental NH_4^+-N load delivery (Figure 10a). The downstream reach of Tuhai-Majia River had a large incremental NH_4^+-N load delivery. The downstream reaches of Daqing River and Luan River had large incremental NH_4^+-N load deliveries.

As shown in Figure 10b, the downstream reach of Yongding River had the highest NH_4^+-N yield. The middle reaches of Beisanhe River and Daqing River had higher NH_4^+-N yield than their downstream reaches. The middle reaches of Zhangweinan Canal and Tuhai-Majia River also had high NH_4^+-N yields.

By combining incremental load, load delivery, and yield results, it could be found that the densely populated area in both PLB and HRB contributed the most to the NH_4^+-N load. However, the difference between the two basins was that the point sources of PLB played the dominant role in the transport of load delivery, whereas the residential land was the dominant sources of load delivery in HRB, as shown in Figure 11. The contributions of farmland in both basins cannot be omitted, either. In PLB, NH_4^+-N load delivery originating from point sources and farmland accounted for 41.83% and 38.84%, respectively. In HRB, NH_4^+-N load transport originating from residential land and farmland accounted for 40.16% and 36.75%, respectively.

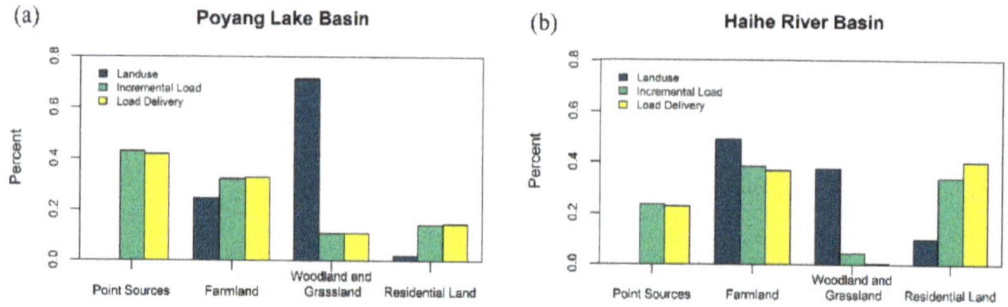

Figure 11. Comparison of land use, incremental load, and load delivery among (**a**) PLB and (**b**) HRB.

The phenomenon above might be explained by the following reasons. Firstly, the proportion of residential land in HRB was much bigger than that in PLB, and thus the residential land contributed more to the load delivered to the outlets in HRB. Secondly, the streamflow of PLB is much larger than that of HRB, which leads to the decrease of NH_4^+-N removal in streamflow and makes the point sources the dominant sources in PLB. Lastly, the urbanization and point sources management of PLB might be inferior to those of HRB.

4.3. *Strategy for Nutrient Management*

Nutrient delivery abatement is vital for the water quality of the receiving waterbody. The results of SPARROW models in this study are meaningful for evaluating NH_4^+-N load transport, critical regions of high load delivery, and dominant nutrient sources.

Based on the studies above, policies of enhancement were proposed. Since PLB has more water resources and steeper terrain, some upstream and middle reaches still transport large amounts of NH_4^+-N load to the outlets, similar to the reaches around the lake. Therefore, it is crucial that more attention should be paid to the reaches around the lake and the upstream and middle reaches, which deliver a large amount of NH_4^+-N load to outlets, especially the centers of big cities. At the same time, the point sources and farmland were recognized as the dominant sources contributing to the load delivered to the outlets in PLB. Consequently, it is important to enhance the management of the point sources in these reaches, such as municipal and industrial wastewater. In addition, since PLB already contains high proportions of woodland and grassland, it is important to establish buffer zones along rivers by planting vegetation or building wetlands, in order to increase the absorption of NH_4^+-N.

Owing to relatively fewer water sources and more plain terrain in HRB, the middle and downstream reaches contributed more to the NH_4^+-N load delivered to the outlets. Hence, the management among these reaches should be enhanced. In HRB, the residential land

and farmland both have critical positions in the delivery of NH_4^+-N loads to the outlets, due to the large amount of residential land and the high population density. Consequently, controlling the domestic pollution and reducing the usage of chemical fertilizers are feasible measures to be undertaken urgently. In addition, increasing the woodland and grassland coverage to enhance the retention of nutrients in land areas may be a sound measure to reduce the NH_4^+-N load delivered to the outlets.

5. Conclusions

The SPARROW model is a spatial explicitly method to address nutrient load and streamflow transport in watersheds. This study developed SPARROW models in two multi-rivers basins in China, which cover large areas in the north and south of China, respectively. These two basins have quite different conditions in many aspects, including weather, water resources, and land use. The SPARROW models were used to evaluate NH_4^+-N load streamflow transport, critical regions of high load delivery, and dominant nutrient sources in these two basins, which further provided basin-specific advice to the authorities. Based on the results of this study, the point sources and the farmland are the dominant source of NH_4^+-N entering Poyang Lake, while the residential land and farmland are the major sources of NH_4^+-N entering Bohai Bay. The following measures should be taken: In PLB, especially among the centers of big cities, it is important to enhance the management of the point sources, such as municipal and industrial wastewater. In addition, it is also advised to establish buffer zones along rivers by planting vegetation or building wetlands, in order to increase the absorption of NH_4^+-N. In HRB, controlling the domestic pollution and reducing the usage of chemical fertilizers are feasible measures that should be urgently considered. Moreover, increasing the woodland and grassland coverage to enhance the retention of nutrients in land areas may be a sound measure for the reduction of the NH_4^+-N load delivered to the outlets.

The SPARROW models built for PLB and HRB can serve as references for future uses for different basins with various conditions, extending this model's scope and adaptability. Remarkably, at twice the size of PLB, HRB has more plain terrain, less water resources and more long-distance canals, which lead to more difficult in its modelling. Hence, the SPARROW model developed in HRB is a worthy possibility for similar research in future.

Supplementary Materials: The following are available online at https://www.mdpi.com/article/10.3390/w14020209/s1, Table S1: Summary of water resources of the main rivers in PLB, Table S2: Summary of water resources of the main rivers in HRB, Table S3: Parameters of GWLF models in PLB.

Author Contributions: Conceptualization, Y.W.; Data curation, H.C.; Formal analysis, J.Y.; Project administration, L.G.; Software, J.Y. and G.L.; Supervision, Y.W.; Visualization, H.C.; Writing—original draft, J.Y.; Writing—review & editing, P.W. All authors have read and agreed to the published version of the manuscript.

Funding: This research was funded by the Major Science and Technology Program for Water Pollution Control and Treatment (2017ZX07301-001).

Institutional Review Board Statement: Not applicable.

Informed Consent Statement: Not applicable.

Data Availability Statement: Publicly available datasets of water resources were analyzed in this study. This data can be found here: Haihe River Water Resources Bulletin 2017, http://www.hwcc.gov.cn/hwcc/static/szygb/gongbao2017/index.html (accessed on 19 November 2021); Jiangxi Water Resources Bulletin 2017, http://slt.jiangxi.gov.cn/resource/uploadfile/file/20180917/20180917112257428.pdf (accessed on 19 November 2021).

Acknowledgments: We wish to thank the Jiangxi Academy of Environmental Sciences for providing streamflow and water quality data of PLB.

Conflicts of Interest: The authors declare no conflict of interest.

References

1. Bouraoui, F.; Grizzetti, B. Modelling mitigation options to reduce diffuse nitrogen water pollution from agriculture. *Sci. Total Environ.* **2014**, *468*, 1267–1277. [CrossRef]
2. Chapra, S.C.; Robertson, A. Great Lakes Eutrophication: The Effect of Point Source Control of Total Phosphorus. *Science* **1977**, *196*, 1448–1450. [CrossRef]
3. Xia, X.; Zhang, S.; Li, S.; Zhang, L.; Wang, G.; Zhang, L.; Wang, J.; Li, Z. The cycle of nitrogen in river systems: Sources, transformation, and flux. *Environ. Sci. Processes Impacts* **2018**, *20*, 863–891. [CrossRef] [PubMed]
4. Manto, M.J.; Xie, P.; Keller, M.A.; Liano, W.E.; Pu, T.; Wang, C. Recovery of ammonium from aqueous solutions using ZSM-5. *Chemosphere* **2018**, *198*, 501–509. [CrossRef] [PubMed]
5. Dai, Y.; Lang, Y.; Wang, T.; Han, X.; Wang, L.; Zhong, J. Modelling the sources and transport of ammonium nitrogen with the SPARROW model: A case study in a karst basin. *J. Hydrol.* **2021**, *592*, 125763. [CrossRef]
6. Jin, L.; Whitehead, P.G.; Heppell, C.M.; Lansdown, K.; Purdie, D.A.; Trimmer, M. Modelling flow and inorganic nitrogen dynamics on the Hampshire Avon: Linking upstream processes to downstream water quality. *Sci. Total Environ.* **2016**, *572*, 1496–1506. [CrossRef]
7. Ervinia, A.; Huang, J.; Zhang, Z. Nitrogen sources, processes, and associated impacts of climate and land-use changes in a coastal China watershed: Insights from the INCA-N model. *Mar. Pollut. Bull.* **2020**, *159*, 111502. [CrossRef]
8. Zhang, Z.; Huang, J.; Xiao, C.; Huang, J.-C. A simulation-based method to develop strategies for nitrogen pollution control in a creek watershed with sparse data. *Environ. Sci. Pollut. Res.* **2020**, *27*, 38849–38860. [CrossRef]
9. Xue, B.; Zhang, H.; Wang, Y.; Tan, Z.; Zhu, Y.; Shrestha, S. Modeling water quantity and quality for a typical agricultural plain basin of northern China by a coupled model. *Sci. Total Environ.* **2021**, *790*, 148139. [CrossRef] [PubMed]
10. Wellen, C.C.; Shatilla, N.J.; Carey, S.K. Regional scale selenium loading associated with surface coal mining, Elk Valley, British Columbia, Canada. *Sci. Total Environ.* **2015**, *532*, 791–802. [CrossRef]
11. Li, X.; Feng, J.; Wellen, C.; Wang, Y. A Bayesian approach of high impaired river reaches identification and total nitrogen load estimation in a sparsely monitored basin. *Environ. Sci. Pollut. Res.* **2017**, *24*, 987–996. [CrossRef] [PubMed]
12. Yuan, L.; Sinshaw, T.; Forshay, K.J. Review of Watershed-Scale Water Quality and Nonpoint Source Pollution Models. *Geosciences* **2020**, *10*, 25. [CrossRef]
13. Smith, R.A.; Schwarz, G.E.; Alexander, R.B. Regional interpretation of water-quality monitoring data. *Water Resour. Res.* **1997**, *33*, 2781–2798. [CrossRef]
14. Alexander, R.B.; Smith, R.A.; Schwarz, G.E. Effect of stream channel size on the delivery of nitrogen to the Gulf of Mexico. *Nature* **2000**, *403*, 758–761. [CrossRef]
15. Benoy, G.A.; Jenkinson, R.W.; Robertson, D.M.; Saad, D.A. Nutrient delivery to Lake Winnipeg from the RedAssiniboine River Basin—A binational application of the SPARROW model. *Can. Water Resour. J.* **2016**, *41*, 429–447. [CrossRef]
16. Brakebill, J.W.; Ator, S.W.; Schwarz, G.E. Sources of Suspended-Sediment Flux in Streams of the Chesapeake Bay Watershed: A Regional Application of the SPARROW Model1. *J. Am. Water Resour. Assoc.* **2010**, *46*, 757–776. [CrossRef]
17. Hoos, A.B.; McMahon, G. Spatial analysis of instream nitrogen loads and factors controlling nitrogen delivery to streams in the southeastern United States using spatially referenced regression on watershed attributes (SPARROW) and regional classification frameworks. *Hydrol. Process.* **2009**, *23*, 2275–2294. [CrossRef]
18. Alexander, R.B.; Schwarz, G.E.; Boyer, E.W. Advances in Quantifying Streamflow Variability Across Continental Scales: 1. Identifying Natural and Anthropogenic Controlling Factors in the USA Using a Spatially Explicit Modeling Method. *Water Resour. Res.* **2019**, *55*, 10893–10917. [CrossRef]
19. Alexander, R.B.; Schwarz, G.E.; Boyer, E.W. Advances in Quantifying Streamflow Variability Across Continental Scales: 2. Improved Model Regionalization and Prediction Uncertainties Using Hierarchical Bayesian Methods. *Water Resour. Res.* **2019**, *55*, 11061–11087. [CrossRef]
20. Ator, S.W.; Garcia, A.M.; Schwarz, G.E.; Blomquist, J.D.; Sekellick, A.J. Toward Explaining Nitrogen and Phosphorus Trends in Chesapeake Bay Tributaries, 1992–2012. *J. Am. Water Resour. Assoc.* **2019**, *55*, 1149–1168. [CrossRef]
21. Brown, J.B.; Sprague, L.A.; Dupree, J.A. Nutrient Sources and Transport in the Missouri River Basin, with Emphasis on the Effects of Irrigation and Reservoirs1. *JAWRA J. Am. Water Resour. Assoc.* **2011**, *47*, 1034–1060. [CrossRef]
22. Rebich, R.A.; Houston, N.A.; Mize, S.V.; Pearson, D.K.; Ging, P.B.; Evan Hornig, C. Sources and Delivery of Nutrients to the Northwestern Gulf of Mexico from Streams in the South-Central United States1. *JAWRA J. Am. Water Resour. Assoc.* **2011**, *47*, 1061–1086. [CrossRef]
23. Puri, D.; Borel, K.; Vance, C.; Karthikeyan, R. Optimization of a Water Quality Monitoring Network Using a Spatially Referenced Water Quality Model and a Genetic Algorithm. *Water* **2017**, *9*, 704. [CrossRef]
24. Domagalski, J.; Saleh, D. Sources and Transport of Phosphorus to Rivers in California and Adjacent States, US, as Determined by SPARROW Modeling. *J. Am. Water Resour. Assoc.* **2015**, *51*, 1463–1486. [CrossRef]
25. Duan, W.L.; He, B.; Takara, K.; Luo, P.P.; Nover, D.; Hu, M.C. Modeling suspended sediment sources and transport in the Ishikari River basin, Japan, using SPARROW. *Hydrol. Earth Syst. Sci.* **2015**, *19*, 1293–1306. [CrossRef]
26. Li, X.; Wellen, C.; Liu, G.; Wang, Y.; Wang, Z.-L. Estimation of nutrient sources and transport using Spatially Referenced Regressions on Watershed Attributes: A case study in Songhuajiang River Basin, China. *Environ. Sci. Pollut. Res.* **2015**, *22*, 6989–7001. [CrossRef] [PubMed]

27. Xu, Z.Z.; Ji, Z.X.; Liang, B.; Song, D.R.; Lin, Y.; Lin, J.G. Estimate of nutrient sources and transport into Bohai Bay in China from a lower plain urban watershed using a SPARROW model. *Environ. Sci. Pollut. Res.* **2021**, *28*, 25733–25747. [CrossRef]
28. Li, G.; Wang, Q.; Liu, G.; Zhao, Y.; Wang, Y.; Peng, S.; Wei, Y.; Wang, J. A Successful Approach of the First Ecological Compensation Demonstration for Crossing Provinces of Downstream and Upstream in China. *Sustainability* **2020**, *12*, 6021. [CrossRef]
29. Zhou, P.; Huang, J.; Hong, H. Modeling nutrient sources, transport and management strategies in a coastal watershed, Southeast China. *Sci. Total Environ.* **2018**, *610*, 1298–1309. [CrossRef] [PubMed]
30. Wang, Y.; Ouyang, W.; Zhang, Y.; Lin, C.; He, M.; Wang, P. Quantify phosphorus transport distinction of different reaches to estuary under long-term anthropogenic perturbation. *Sci. Total Environ.* **2021**, *780*, 146647. [CrossRef]
31. Elliott, A.H.; Alexander, R.B.; Schwarz, G.E.; Shankar, U.; Sukias, J.P.S.; McBride, G.B. Estimation of nutrient sources and transport for New Zealand using the hybrid mechanistic-statistical model SPARROW. *J. Hydrol.–N. Z.* **2005**, *44*, 1–27.
32. Alexander, R.B.; Elliott, A.H.; Shankar, U.; McBride, G.B. Estimating the sources and transport of nutrients in the Waikato River Basin, New Zealand. *Water Resour. Res.* **2002**, *38*, 4-1–4-23. [CrossRef]
33. Aguilera, R.; Marce, R.; Sabater, S. Linking in-stream nutrient flux to land use and inter-annual hydrological variability at the watershed scale. *Sci. Total Environ.* **2012**, *440*, 72–81. [CrossRef]
34. Aguilera, R.; Marce, R.; Sabater, S. Modeling nutrient retention at the watershed scale: Does small stream research apply to the whole river network? *J. Geophys. Res. Biogeosci.* **2013**, *118*, 728–740. [CrossRef]
35. Miller, M.P.; de Souza, M.L.; Alexander, R.B.; Sanisaca, L.G.; Teixeira, A.d.A.; Appling, A.P. Application of the RSPARROW Modeling Tool to Estimate Total Nitrogen Sources to Streams and Evaluate Source Reduction Management Scenarios in the Grande River Basin, Brazil. *Water* **2020**, *12*, 2911. [CrossRef]
36. Zhang, W.; Li, H.; Kendall, A.D.; Hyndman, D.W.; Diao, Y.; Geng, J.; Pang, J. Nitrogen transport and retention in a headwater catchment with dense distributions of lowland ponds. *Sci. Total Environ.* **2019**, *683*, 37–48. [CrossRef] [PubMed]
37. Robertson, D.M.; Saad, D.A. Nutrient Inputs to the Laurentian Great Lakes by Source and Watershed Estimated Using SPARROW Watershed Models1. *JAWRA J. Am. Water Resour. Assoc.* **2011**, *47*, 1011–1033. [CrossRef]
38. Garcia, A.M.; Alexander, R.B.; Arnold, J.G.; Norfleet, L.; White, M.J.; Robertson, D.M.; Schwarz, G. Regional Effects of Agricultural Conservation Practices on Nutrient Transport in the Upper Mississippi River Basin. *Environ. Sci. Technol.* **2016**, *50*, 6991–7000. [CrossRef]
39. Zhang, W.; Pueppke, S.G.; Li, H.; Geng, J.; Diao, Y.; Hyndman, D.W. Modeling phosphorus sources and transport in a headwater catchment with rapid agricultural expansion. *Environ. Pollut.* **2019**, *255*, 113273. [CrossRef]
40. Robertson, D.M.; Saad, D.A.; Christiansen, D.E.; Lorenz, D.J. Simulated impacts of climate change on phosphorus loading to Lake Michigan. *J. Great Lakes Res.* **2016**, *42*, 536–548. [CrossRef]
41. Alam, M.J.; Goodall, J.L.; Bowes, B.D.; Girvetz, E.H. The Impact of Projected Climate Change Scenarios on Nitrogen Yield at a Regional Scale for the Contiguous United States. *J. Am. Water Resour. Assoc.* **2017**, *53*, 854–870. [CrossRef]
42. Morales-Marin, L.; Wheater, H.; Lindenschmidt, K.-E. Potential Changes of Annual-Averaged Nutrient Export in the South Saskatchewan River Basin under Climate and Land-Use Change Scenarios. *Water* **2018**, *10*, 1438. [CrossRef]
43. Miller, M.P.; Capel, P.D.; Garcia, A.M.; Ator, S.W. Response of Nitrogen Loading to the Chesapeake Bay to Source Reduction and Land Use Change Scenarios: A SPARROW-Informed Analysis. *J. Am. Water Resour. Assoc.* **2020**, *56*, 100–112. [CrossRef]
44. Zhang, Z.; Chen, X.; Xu, C.-Y.; Hong, Y.; Hardy, J.; Sun, Z. Examining the influence of river-lake interaction on the drought and water resources in the Poyang Lake basin. *J. Hydrol.* **2015**, *522*, 510–521. [CrossRef]
45. Tang, W.; Zhang, W.; Zhao, Y.; Zhang, H.; Shan, B. Basin-scale comprehensive assessment of cadmium pollution, risk, and toxicity in riverine sediments of the Haihe Basin in north China. *Ecol. Indic.* **2017**, *81*, 295–301. [CrossRef]
46. Zhao, G.; Hoermann, G.; Fohrer, N.; Zhang, Z.; Zhai, J. Streamflow Trends and Climate Variability Impacts in Poyang Lake Basin, China. *Water Resour. Manag.* **2010**, *24*, 689–706. [CrossRef]
47. Li, F.; Zhan, C.; Xu, Z.; Jiang, S.; Xiong, J. Remote sensing monitoring on regional crop water productivity in the Haihe River Basin. *J. Geogr. Sci.* **2013**, *23*, 1080–1090. [CrossRef]
48. Morales-Marin, L.A.; Wheater, H.S.; Lindenschmidt, K.E. Assessment of nutrient loadings of a large multipurpose prairie reservoir. *J. Hydrol.* **2017**, *550*, 166–185. [CrossRef]
49. Haith, D.A.; Shoemaker, L.L. GENERALIZED WATERSHED LOADING FUNCTIONS FOR STREAM FLOW NUTRIENTS1. *JAWRA J. Am. Water Resour. Assoc.* **1987**, *23*, 471–478. [CrossRef]
50. Sha, J.; Swaney, D.R.; Hong, B.; Wang, J.; Wang, Y.; Wang, Z.-L. Estimation of watershed hydrologic processes in arid conditions with a modified watershed model. *J. Hydrol.* **2014**, *519*, 3550–3556. [CrossRef]
51. Schwarz, G.; Hoos, A.; Alexander, R.; Smith, R. *The SPARROW Surface Water-Quality Model: Theory, Application and User Documentation*; U.S. Geological Survey Techniques and Methods: Reston, VA, USA, 2006.
52. Qi, Z.D.; Kang, G.L.; Chu, C.L.; Qiu, Y.; Xu, Z.; Wang, Y.Q. Comparison of SWAT and GWLF Model Simulation Performance in Humid South and Semi-Arid North of China. *Water* **2017**, *9*, 567. [CrossRef]
53. Liu, X.; Wang, Y.; Feng, J.; Chu, C.; Qiu, Y.; Xu, Z.; Li, Z.; Wang, Y. A Bayesian modeling approach for phosphorus load apportionment in a reservoir with high water transfer disturbance. *Environ. Sci. Pollut. Res.* **2018**, *25*, 32395–32408. [CrossRef] [PubMed]
54. Schwarz, G.E. *A Preliminary SPARROW Model of Suspended Sediment for the Conterminous United States*; U.S. Geological Survey Open-File Report 2008–1205; U.S. Geological Survey: Reston, VA, USA, 2008; p. 8.

55. Saleh, D.; Domagalski, J. SPARROW Modeling of Nitrogen Sources and Transport in Rivers and Streams of California and Adjacent States, US. *J. Am. Water Resour. Assoc.* **2015**, *51*, 1487–1507. [CrossRef]
56. Morales-Marin, L.A.; Wheater, H.S.; Lindenschmidt, K.E. Estimating Sediment Loadings in the South Saskatchewan River Catchment. *Water Resour. Manag.* **2018**, *32*, 769–783. [CrossRef]
57. Detenbeck, N.E.; You, M.; Torre, D. Recent Changes in Nitrogen Sources and Load Components to Estuaries of the Contiguous United States. *Estuaries Coasts* **2019**, *42*, 2096–2113. [CrossRef]

water

MDPI

Article

Proof-of-Concept of a Quasi-2D Water-Quality Modelling Approach to Simulate Transverse Mixing in Rivers

Pouya Sabokruhie [1], Eric Akomeah [2], Tammy Rosner [3] and Karl-Erich Lindenschmidt [1,*]

[1] Global Institute for Water Security, School of Environment and Sustainability, University of Saskatchewan, 11 Innovation Boulevard, Saskatoon, SK S7N 3H5, Canada; p.sabokruhie@usask.ca
[2] Water Security Agency, 111 Fairford Street East, Moose Jaw, SK S6H 7X9, Canada; eric.akomeah@wsask.ca
[3] Four Elements Consulting Ltd., 1012 18th Ave. SE, Calgary, AB T2G 1L6, Canada; tammy.rosner@Fourelementsconsulting.ca
* Correspondence: karl-erich.lindenschmidt@usask.ca; Tel.: +1-(306)-966-6174

Abstract: A quasi-two-dimensional (quasi-2D) modelling approach is introduced to mimic transverse mixing of an inflow into a river from one of its banks, either an industrial outfall or a tributary. The concentrations of determinands in the inflow vary greatly from those in the river, leading to very long mixing lengths in the river downstream of the inflow location. Ideally, a two-dimensional (2D) model would be used on a small scale to capture the mixing of the two flow streams. However, for large-scale applications of several hundreds of kilometres of river length, such an approach demands too many computational resources and too much computational time, especially if the application will at some point require ensemble input from climate-change scenario data. However, a one-dimensional (1D) model with variables varying in the longitudinal flow direction but averaged across the cross-sections is too simple of an approach to capture the lateral mixing between different flow streams within the river. Hence, a quasi-2D method is proposed in which a simplified 1D solver is still applied but the discretisation of the model setup can be carried out in such a way as to enable a 2D representation of the model domain. The quasi-2D model setup also allows secondary channels and side lakes in floodplains to be incorporated into the discretisation. To show proof-of-concept, the approach has been tested on a stretch of the lower Athabasca River in Canada flowing through the oil sands region between Fort McMurray and Fort MacKay. A dye tracer and suspended sediments are the constituents modelled in this test case.

Keywords: lower Athabasca River; oil sands region; quasi-2D modelling; water-quality analysis simulation program (WASP); water-quality modelling

Citation: Sabokruhie, P.; Akomeah, E.; Rosner, T.; Lindenschmidt, K.-E. Proof-of-Concept of a Quasi-2D Water-Quality Modelling Approach to Simulate Transverse Mixing in Rivers. *Water* **2021**, *13*, 3071. https://doi.org/10.3390/w13213071

Academic Editors: Leon Boegman and Tammo Steenhuis

Received: 16 September 2021
Accepted: 30 October 2021
Published: 2 November 2021

1. Introduction

1.1. Background

Much of the process water from oil sands surface mining operations is recycled and managed in tailing ponds. However, the capacity for storage is approaching unmanageable and unsustainable levels; hence, some release of treated process water into the Athabasca River is anticipated as early as 2025 [1]. The release of treated process water may pose a risk to aquatic species and to humans who harvest and consume these species, in particular fish. Therefore, effective models to describe the transport and fate of oil sands related substances are required [2]. These substances can be transported downstream and deposited in Lake Athabasca and the Peace-Athabasca-Delta (PAD); in addition, secondary channels and lakes within the floodplain along the lower river reach may also trap released sediment and associated constituents. An important objective of this research is to determine the fate of such effluent within these features using computer modelling.

1.2. Water-Quality Modelling

The development of complex three-dimensional integrated hydrodynamic, sediment transport, and water quality models has been proposed to characterize the transport and fate of sediment and associated constituents in the lower Athabasca River, advocated by several researchers and government agencies [2]. However, the implementation of complex modelling frameworks may not be advantageous for many reasons, including cost, the time required to develop the framework, and lengthy model simulation times. Additionally, model complexity can obfuscate rather than elucidate key processes, and both overly complex and overly simple models can have reduced reliability.

We propose using the Water Quality Analysis Simulation Program (WASP) [3] to develop a model to characterize the fate and transport of sediment and associated constituents within the lower Athabasca River. WASP is a dynamic compartment-modelling program for aquatic systems, including both the water column and the underlying benthos. Its representation of sediment and material kinetics is more sophisticated than other commonly used models [4]. WASP is a widely used framework for developing site-specific models for simulating toxicant concentrations in surface waters and sediments over a range of complexities and temporal and spatial scales. WASP has an advanced toxicant module that includes representation of a range of solids classes, with individual physical and chemical characteristics. Solids classes can be organic materials (e.g., plankton, algae, detritus) or inorganic (e.g., sand, silt, clay).

1.3. Quasi-Two-Dimensional Modelling

This paper describes a unique quasi-two-dimensional representation of river hydraulics that is particularly suited to the application of the WASP model to accurately represent the configuration of the lower Athabasca River with a high level of computational efficiency. In the current study, we introduce a novel approach to modelling transverse mixing in a river with secondary channels and side lakes to study the water quality along the area of the Athabasca River with extensive oil sands development.

In order to maintain short computational times, a one-dimensional (1D) modelling approach is necessary. However, the transverse mixing along the river requires modelling with at least a two-dimensional (2D) representation, especially since the lengths of complete mixing along this river are relatively long (>100 km). Hence, the use of a quasi-two-dimensional (quasi-2D) approach is proposed, in which flow is simulated in 1D, but in such a way to allow a 2D discretisation of the domain.

Quasi-2D water-quality modelling has been carried out in the past for other applications. For off-channel storage facilities (polders) along the Elbe River in Germany, Lindenschmidt et al. [5] modelled dissolved oxygen and nutrient dynamics for various flow regimes (low, medium, high (flood) flows). Deposition of sediments and heavy metals in the off-channel storage basins was captured using the quasi-2D method [6–8]. The quasi-2D approach has also been used to capture flows between main river channels and their floodplains, in particular through dike breaches [9] and capping flood peaks using side-channel storage [10,11]. Flow between the Mekong River and its delta [12] and between the Po River and a portion of its floodplain [13] were simulated using quasi-2D models. Sediment transport was included in a quasi-2D model of the Rhine River's main channel and floodplain [14]. In this study, we extend the quasi-2D approach to model transverse mixing in rivers.

1.4. Objectives

The objectives of this study are to:

1. Develop a quasi-2D modelling approach to characterise transverse mixing of two water streams of different sediment concentrations.
2. Assess the role of drawing on the output of additional models to complement the implementation of models with different complexity and spatial scale.

3. As proof of concept of the approach, the sediment transport along the lower Athabasca River was quantified. The model domain includes a secondary channel and side lake connected hydraulically to the river's main stem.

2. Site Description

The Athabasca River (see Figure 1) is a northern river (ice-covered in winter) in western Canada that originates on the eastern slopes of the Rocky Mountains and flows approximately 1500 km in a north-easterly direction to empty into Lake Athabasca. It is the longest unregulated river in Alberta. For the last 80 km, the river flows through the Athabasca Delta. The Athabasca Delta is part of the larger Peace-Athabasca Delta (PAD), which is the largest inland delta in North America. The PAD is an important ecosystem for many aquatic and terrestrial plant and animal species, and the area has been named both a Ramsar and United Nations Educational, Scientific and Cultural Organization (UNESCO) World Heritage site. From a human perspective, the PAD is an important social, cultural and economic entity for many of the Aboriginal communities in the area. Many conservation efforts have been carried out to maintain the ecological integrity of the aquatic and terrestrial systems.

Referring to Figure 1, the average discharge at the Water Survey of Canada (WSC) gauge at Embarras (gauge 07DD001–Athabasca River at Embarras Airport), which is approximately 80 km upstream of the PAD, is around 850 m^3/s. The total catchment area draining at the same gauge is approximately 156,000 km^2. Much of the basin's land use consists of forests (89%) with some agricultural lands (8%). The basin is sparsely populated (≈200,000 inhabitants), with the population concentrated in a few urban centres, which release effluent from municipal wastewater treatment plants into the river.

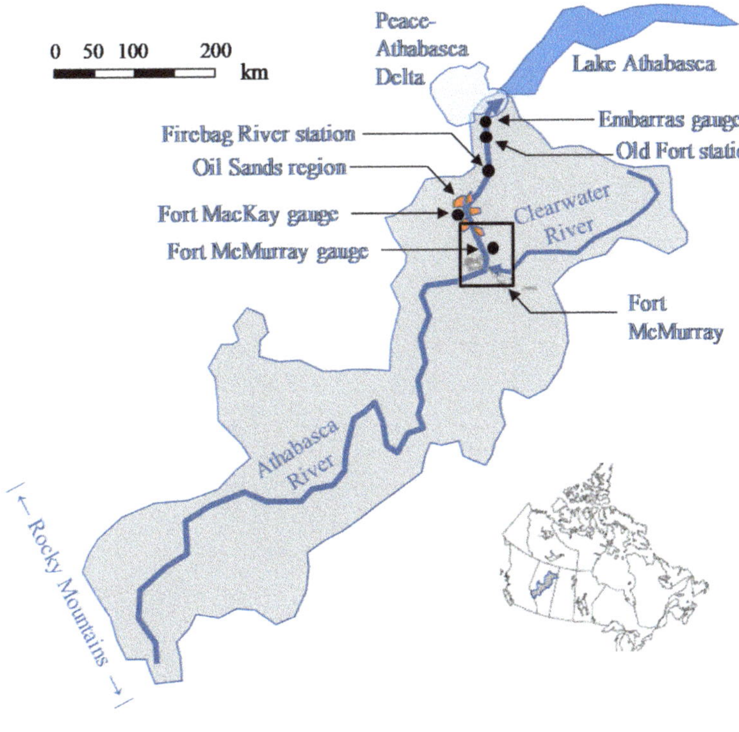

Figure 1. Athabasca River basin. Inset corresponds to the realm shown in Figure 2.

The oil sand region covers a large area of Northern Alberta and has one of the richest deposits of petroleum in the world. For a 65 km stretch of the lower Athabasca River, between Fort McMurray and downstream of Fort MacKay (see Figure 1), the river flows through the active oil sand surface mines. Much of the oil is extracted through open-pit mining and processed using water from the Athabasca River. Almost 5% of the river's average flow has been allocated for anthropogenic usage. Although approximately half of that amount is allocated for surface and in situ mining activities, less than 1% is used [15].

The lower Athabasca River contains many islands, secondary channels, wetlands, and floodplain lakes. Many of these secondary channels freeze to the bottom or have low oxygen levels during winter [16]. Sediment deposition areas occur downstream of tributaries (confluence bars), in mid-channel bars, in secondary channels and in side lake and wetland areas along the river. During high flow events, substantial amounts of oil sands material can be transported from tributaries into the river to be deposited in these depositional areas. Over time, these areas of high oil sands material are mixed and diluted with water and sediment coming from upstream.

To show proof-of-concept, the domain modelled in this study is only a short stretch of 15 km. It is part of a larger system (>200 km) that is to be modelled in the future, with additional outfalls and sensitive depositional areas (e.g., secondary channels and side lakes). The model domain is indicated in Figure 2 and was chosen based on the location of proposed outfalls from oilsands operations. One particular side lake of interest in this study is Shipyard Lake, which is located near the Suncor oil facilities on the east side of the Athabasca River and is within the model domain. The lake, shown in Figure 3, has a surface area of 21.3 ha and a maximum depth of 2.2 m [17]. It is essentially a large wetland area that is flooded by higher flows from the Athabasca River [17]. Numerous small creeks feed the lake and water from the Athabasca River flows into Shipyard Lake over a low levee between the watercourse and water body. Flow begins to overflow the levee at an Athabasca River flow of 2800 m^3/s [18], which corresponds to a water surface elevation of 237.7 m a.s.l. just upstream of the lake [18]. This corresponds to a water level elevation of approximately 240.1 m a.s.l. at the Athabasca River gauge downstream of Fort McMurray [19], the location of which is shown in Figure 2.

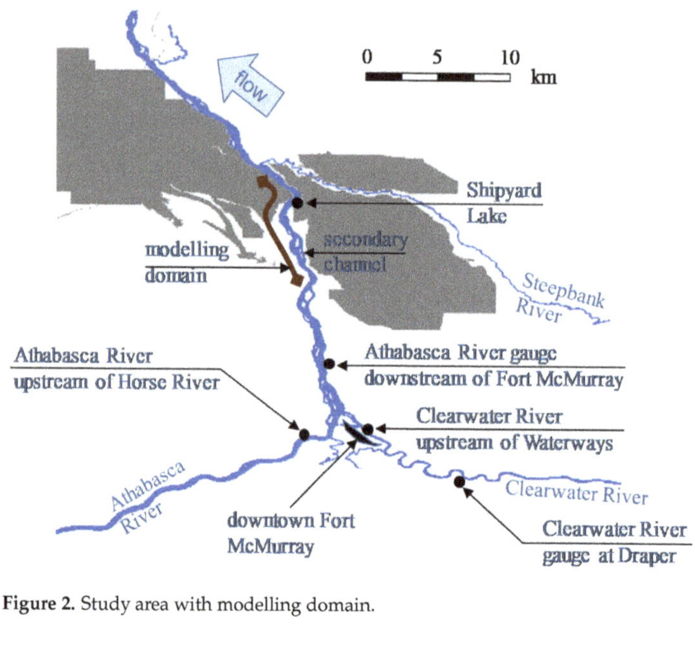

Figure 2. Study area with modelling domain.

Figure 3. Shipyard Lake (**left** panel) and secondary channel (**right** panel). Both Google Earth maps are at the same scale.

A secondary channel is also included in the model domain, also shown in Figure 3. Indicated in the figure is the location of a cross-section of the Athabasca River main stem and the secondary channel, which is shown in Figure 4. Some water does flow into the secondary channel from the Athabasca River at a flow of 600 m^3/s.

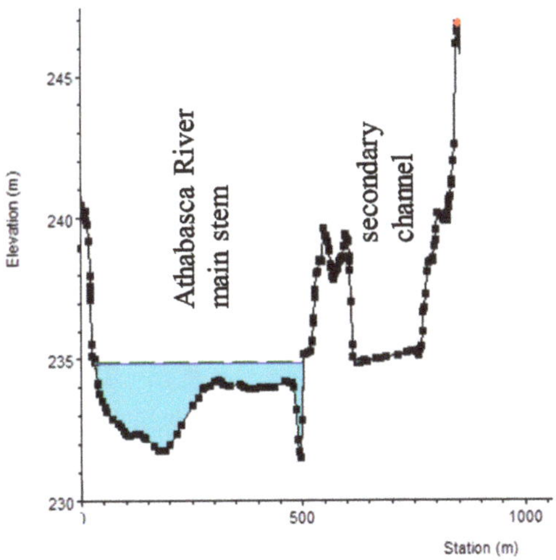

Figure 4. Hydrologic Engineering Center-River Analysis System (HEC-RAS) cross-section of the Athabasca River main stem and secondary channel with water-level profile at a flow of 600 m^3/s. The location of the cross-section in indicated in Figure 3.

A major tributary of the Athabasca River is the Clearwater River (see Figures 1 and 2), which flows from the east into the Athabasca River at Fort McMurray. For an average flow of 600 m^3/s at the gauge on the Athabasca River downstream of Fort McMurray (gauge 07DA001–Athabasca River below Fort McMurray), the average annual flow from the Clearwater River recorded at the Draper gauge (gauge 07CD001–Clearwater River at Draper) is 116 m^3/s and ranges from 30 to 500 m^3/s (see Figure 5).

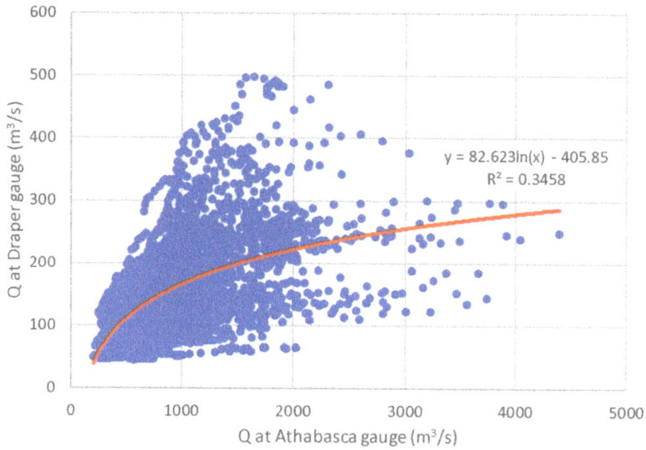

Figure 5. Flows recorded by the Athabasca River gauge below Fort McMurray (gauge 07DA001) and the Clearwater River gauge at Draper (gauge 07CD001). Flows are of daily means recorded on the same day over a 30-year period (1990–2019)).

Total suspended solids (TSS) data from four stations along the Athabasca River were used in the analysis:

- Athabasca River upstream of the Firebag River confluence (Alberta Environment and Park's long term monitoring station AB07DA0980; Figure 1)
- Old Fort/Devil's Elbow (Alberta Environment and Park's long term monitoring station AB07DD0010 (open water) and AB07DD0105 (ice-cover); Figure 1)
- Athabasca River upstream of the Horse River—(Alberta Environment and Park's long term monitoring station AB07CC0030; Figure 2)
- Clearwater River upstream of Waterways (Alberta Environment and Park's monitoring station AB07CD0200/210/090 Clearwater River upstream of Waterways; Figure 2)

Empirical relationships between TSS concentrations and river discharge are provided in Appendix A, from which equations were established and are shown in Figure 6 for a lower flow regime. Sediment transport was simulated only for the 2800 m^3/s flow since, in this scenario, Athabasca River water flows into both the secondary channel and Shipyard Lake. A total flow of 2800 m^3/s at the Athabasca River gauge below Fort McMurray corresponds to a flow of 250 m^3/s for the Clearwater River (from Figure 5), yielding, through simple subtraction, a flow of 2550 m^3/s at the Athabasca River station upstream of Fort McMurray. From Figure 6, the TSS concentrations are respectively 575.5 and 86.8 mg/L for the upper Athabasca and Clearwater rivers.

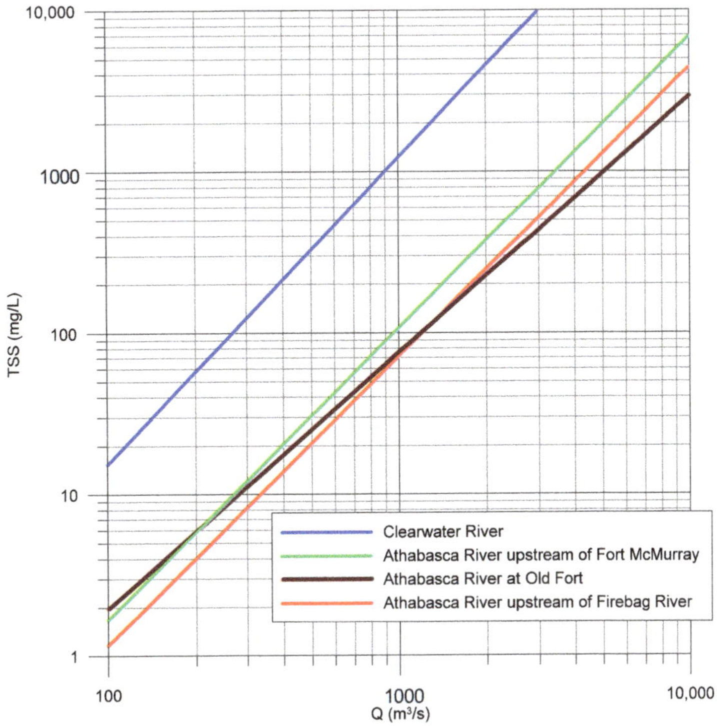

Figure 6. Discharge (daily means) vs. total suspended sediment concentrations.

3. Hydraulic Modelling Setup

Water level gauges are not present within the 15 km modelling domain. There is an active gauge (records from 1957 to present) at the Water Survey of Canada (WSC) station approximately 13 km upstream of the modelling domain at Fort McMurray (gauge 07DA001–Athabasca River below Fort McMurray) and a discontinued gauge (recordings only from 1959 to 1966) approximately 16 km downstream of the modelling domain at Fort MacKay (gauge 07DA003–Athabasca River near Fort MacKay). The average flow at the gauge at Fort McMurray is about 600 m^3/s. This flow was used to calibrate the WASP simulations of tracer mixing from a hypothetical outfall from the right bank at the upstream boundary of the model domain. A flow of 2800 m^3/s was used for validation and to simulate flow from the Athabasca River into Shipyard Lake.

The hydraulic model HEC-RAS [20], developed by the U.S Army Corps of Engineers, was used to generate the segmentation for the WASP water quality model. The water level elevations of the HEC-RAS model at flows 600 and 2800 m^3/s, which formed the geometrical basis for the WASP segmentation, were verified with flow simulations using the hydraulic model ONE-D, a hydrodynamic model that uses a finite difference implicit scheme for the solution ([21,22] both referenced in [23]). Both mass and momentum of sub-critical fluid motion are conserved for open-channel flow. The model is robust and well-tested with many consultancy applications (e.g., [24,25]). Bathymetry data were obtained from Alberta Environment and Parks (AEP) [26].

Figure 7 shows how the water level profiles modelled in ONE-D match with the water level elevations recorded at the gauges at Fort McMurray and Fort MacKay for an average flow of 600 m^3/s and a higher flow of 2800 m^3/s. The water level elevations modelled with HEC-RAS for the same steady-state flow coincide well with the ONE-D water level profile.

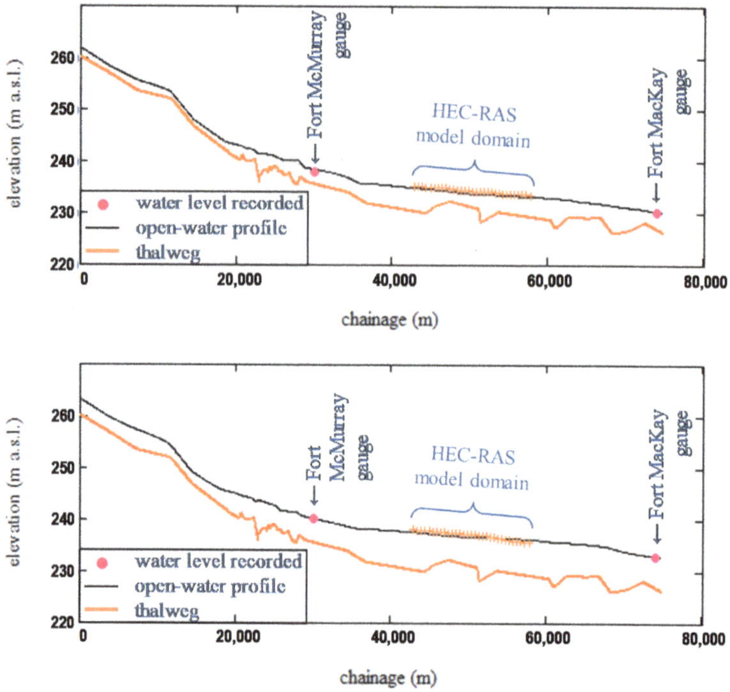

Figure 7. Verification of water surface elevations of the HEC-RAS model with ONE-D for flows of 600 m^3/s (**top** panel) and 2800 m^3/s (**bottom** panel).

4. Water-Quality Modelling Setup

The Water Quality Analysis Simulation Program (WASP 8.32) was used to model the water quality of the river-lake system. WASP is a general dynamic model that applies a segmentation network to solve the conservation of momentum, energy, and mass equations and simulate contaminant and sediment transport. WASP was initially developed in the 1980s and has undergone many upgrades since then. WASP is a widely used model, especially in North America, for addressing various environmental and water quality concerns. The WASP segmentation can be structured in 1D for streams, or 2D for rivers with branching channels or 3D for lakes.

The WASP stream transport module, TOXI, is able to calculate the flow of water, sediment and dissolved constituents through branched and ponded segments and is coupled with flow routing for free flow streams, ponded segments, and backwater reaches. The kinematic wave was used for the 1D flow routing, which is based on solutions of one-dimensional continuity equations and a simplified form of the momentum equation that considers effects of gravity and friction, and calculates variations in velocities, widths and depths throughout the network. Flow through the ponded segments is calculated based on a sharp-crested weir equation, which calculates outflow based on water elevation above the weir crest, using kinematic wave flow for a river and ponded weir overflow for a lake. For the current study, boundary and initial conditions were defined based on water quality data of the closest station to the study area by spatial linear interpolation.

Input data for the model includes channel geometry, flow routing, boundary conditions, environmental time functions, loads and initial conditions for segments. Geometry for the WASP segments (average depths, widths and slopes) was extracted from the fifty-eight HEC-RAS cross-sections, which were spaced at a distance of 493 m apart. Segments are a series of discretized components used to estimate the model state variables at every time step. The sediment transport model is an independent set of routines with associated parameters for the model.

The domain was modelled with a network of river segments, a secondary channel and a side lake. Water quality modelling of such a complex system requires special consideration of the varying hydrodynamics within the system. The primary segments of the system are 493 m long; as mentioned before, right stream segments accommodate 20% of the width of cross-sections produced by HEC-RAS for the 600 m^3/s scenario and 10% of the width for the 2800 m^3/s flow scenario.

Bathymetric information for Shipyard Lake was obtained from a bathymetry survey carried out by Golder [17]. The lake was divided into two segments:

- a deep part with a surface area of 14 ha and an average depth of 1.5 m and
- a shallow part with a surface area of 7.3 ha and an average depth of 0.75 m.

As mentioned above, the mixing length for the stream is long, and the geographical complexity of the study area cannot be represented with a 1D modelling approach alone. On the other hand, a 2D approach would require higher data computational resources for such a long stream. Quasi-2D models can be a reasonable alternative by using the simplicity of a 1D solution but still capturing 2D discretisation of the substance transport. The approach is purely a balance of water and mass between each adjoining segments. This includes the exchange of water and mass between each adjacent-lying, left and right stream segments. The exchange is a small percentage of the water flowing into each of those segments from their upstream segments. The exchange alternates from right-to-left and left-to-right stream segments along the model domain. The flow exchange was calibrated in such a way that the constituents within the two streams mixed together more and more as we moved further downstream along the domain. The flow transfers between adjacent segments were calibrated so that the concentrations matched the observed dispersion pattern calculated with the Athabasca River Model (ARM). Concentration gradients were generally higher between the left and right stream segments at the upstream boundary leading to more rapid mixing in this area; as we move further downstream, the mixing between the adjacent segments of each stream led to a decrease in their concentration gradients and a decrease in the mixing rate.

In this study, results from one model, ARM, were used to calibrate the mixing simulated with WASP. However, field data sampled not only longitudinally along the flow direction of the river, but also transversely across the river can be used to calibrate the model, an example of which is provided in a follow-up study to simulate the transport of vanadium [27].

The stretches between every two cross-sections of the HEC-RAS model were divided into two stream tubes. The flow from the Clearwater River controlled the segment on the right side (right stream), and the main Athabasca River system controlled the segments on the left side (left stream), as shown in Figure 8. Flow for the secondary channel was supplied through the right side stream. It was estimated that 2% of flow into Shipyard Lake was required to maintain water volume continuity. The flow through Shipyard Lake was anticipated to flow back into the right side stream just downstream of the lake. A schematic of the segmentation is shown in Figure 9.

Figure 8. Discretising left and right stream along the Athabasca River modelling domain (map source: Google Earth).

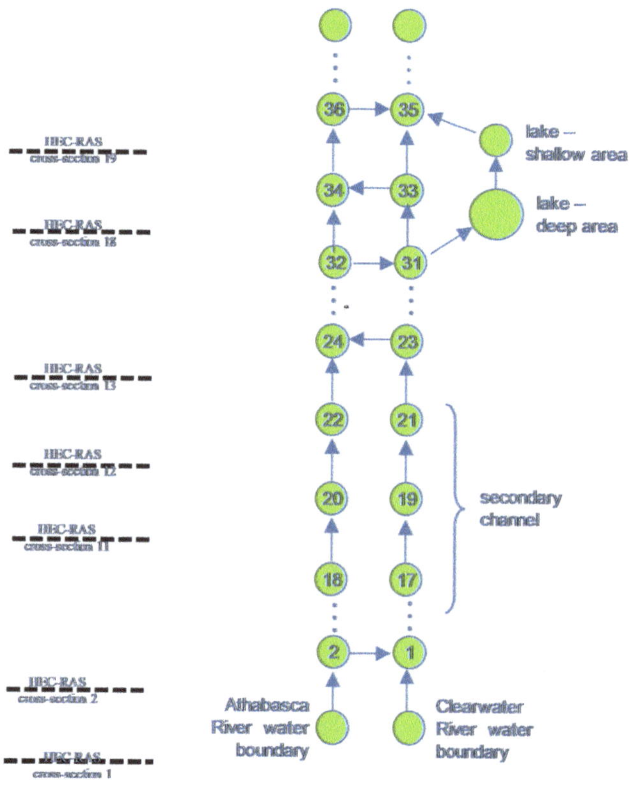

Figure 9. Segmentation of the model domain. left panel: HEC-RAS cross sections; right panel: WASP surface segments.

In order to characterize erosion and sedimentation, benthic segments were added under each surface water segment. Solids can be separated into clay, silt and sand, which can refine simulation results of solids. The user has four choices for simulating solids in WASP: solid flow fields, descriptive, the mechanistic Van Rijn equation, and the mechanistic Robert equation. Currently, WASP has two kinetic modules for modelling sediment transport. For the current study, the advanced toxicant module was used because it is considered to be the best module to simulate oil-sands associated substances.

5. WASP Mixing Calibration (Q = 600 m^3/s)

Calibration of the mixing of the WASP model was done using simulation results from the Athabasca River Model (ARM) [16,17]. ARM is a vertically averaged, two-dimensional mixing model used to predict substance concentrations varying across the width and length of the lower Athabasca River for various flow conditions. The model's calculations are analytical/explicit in nature, so there is no discretization other than differentiating between different reaches with different mixing and hydraulic parameters. ARM applies analytical solutions for river dispersion equations under steady-state conditions. Mid-field mixing is represented by dispersion equations described by Fischer et al. [28]. The same representation is used for passive ambient mixing in the Cornel Mixing Zone Expert System (CORMIX) [29] except that, for ARM, transverse mixing coefficients have been calibrated based on a tracer-dye study completed in 2003 [30,31] and validated using a range of data sources [32]. The calibration of ARM is described in [30,31].

A hypothetical conservative tracer was chosen to calibrate the mixing path of the WASP model to the ARM predictions. A load of 5670 kg/day, which led to the chemical concentration of 1 mg/l in the most-upstream segment (Segment 1) was added. The flow system included two upstream boundary conditions: the Athabasca River water for the left stream boundary and the Clearwater River water for the right stream boundary.

As expected, the conservative tracer was transported to the left stream when a flow was designated between them. Segments divided by islands (segment combinations 16–17, 18–19 and 20–21) had no flow between them. WASP simulated the concentrations of the tracer in each segment. These values were compared with values predicted by the ARM model and shown in Figure 10.

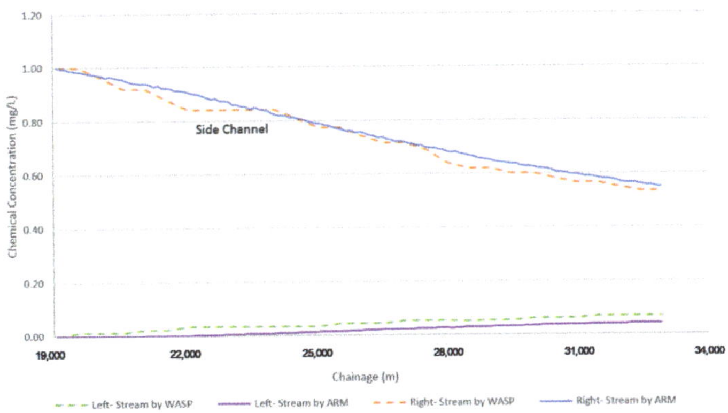

Figure 10. Calibration of WASP with ARM data of the mixing between the left and right streams of the Athabasca River at a flow of 600 m^3/s.

Good agreement for both left and right streams was observed. As can be seen, both models predicted the same trend for both streams: a reduction in chemical concentrations in the right stream and an increase in chemical concentrations in the left stream. Contour plots of the tracer concentrations modelled in WASP and ARM are shown in Figure 11.

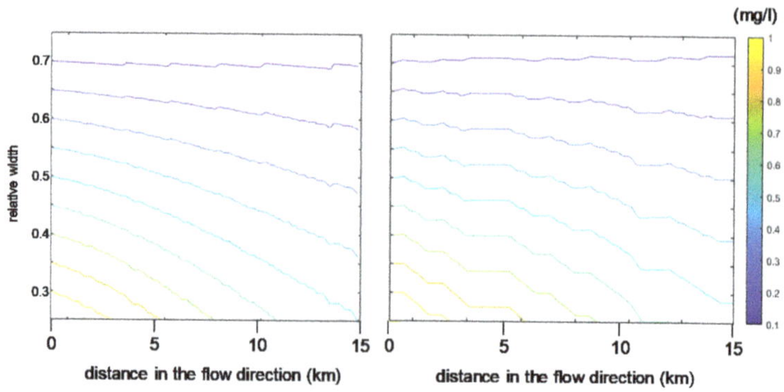

Figure 11. Contour maps for tracer concentration based on the ARM model (**left** panel) and WASP model (**right** panel) for the 600 m^3/s flow scenario. Relative width is based on the fraction of the river width.

6. WASP Mixing Validation (Q = 2800 m^3/s)

A simulation with a higher flow of 2800 m^3/s was carried out for which both Shipyard Lake and the secondary channel became flooded. The results of the two models for the two streams are shown in Figure 12. Again, mixing is evident since the tracer concentrations are increasing in the right stream and decreasing in the left stream, both in the flow direction. Mixing is more rapid for the 2800 m^3/s scenario since the tracer concentrations drop to 40% of the original concentration at the end of the 15 km stretch, whereas the concentrations only drop to 55% in the 600 m^3/s scenario (Figure 10). Contours of the tracer concentrations simulated in ARM and WASP are provided for the 2800 m^3/s flow case in Figure 13. The same trend for the mixing was observed for both flow rates studied herein. For the higher flow rate, faster mixing was observed, which can be explained by higher turbulence and more chaotic eddies which follow observations by Dimotakis [33] and Sreenivasan [34].

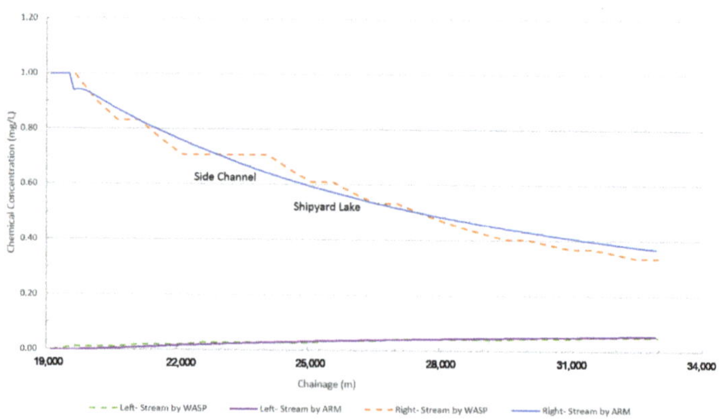

Figure 12. Validation of WASP with ARM of the mixing between the left and right streams of the Athabasca River at a flow of 2800 m^3/s.

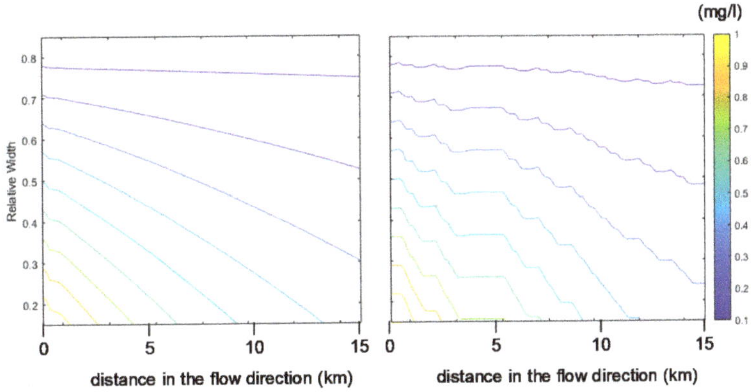

Figure 13. Contour maps for tracer concentration based on the ARM model (**left** panel) and WASP model (**right** panel) for 2800 m^3/s flow scenario. Relative width is based on the fraction of the river width.

7. Sediment Transport Modelling

Suspended solids and benthic sediments are key components of the water quality of rivers and lakes. When the flow velocity decreases, as is the case when, for example, a river flows into a lake or reservoir, suspended sediments will settle out. Key constituents in oil sands process water, such as selenium and polycyclic aromatic hydrocarbons, tend to sorb to fine sediments. For the current study, separate size classes were represented using the descriptive solids transport option in WASP. For the descriptive solids transport, constant settling and resuspension velocities were defined for each segment. Settling velocity of silt and clay were calculated from Stokes Law, and the settling velocity of sand was estimated based on the equation from Ferguson and Church [35].

Size classes for suspended solids were estimated based on suspended sediment samples collected from the Lower Athabasca River using continuous flow centrifugation in 2012 and 2013 as part of the Joint Oil Sands Monitoring Program [1].

Measured TSS data were fitted versus the flow rate for the sediment sampling stations. Power functions were used to fit the data yielding acceptable r^2 values ($r^2 > 0.6$) for all correlations. This provided equations to estimate TSS based on the given flow rates. For the sediment transport simulation, we used the flow scenario of 2800 m^3/s because at that flow, both the secondary channel and Shipyard Lake are flooded by Athabasca River water. A flow of 2800 m^3/s corresponds to a flow of 250 m^3/s for the Clearwater River (from Figure 5), and a flow of 2550 m^3/s along the upper Athabasca River reach, which represents a flow ratio of approximately 10–90%, respectively. Although the model domain is situated downstream of the Athabasca/Clearwater river confluence, for the purpose of showing proof-of-concept in this study, the 10/90 percentage ratio was maintained for the right/left streams.

After obtaining TSS values for the flow scenario for the water quality stations, an interpolation was done to obtain initial sediment concentrations for each segment along the study area (see Figure 14). For segments on the right stream (segments with odd numbers), linear interpolation between Clearwater River at Draper and Athabasca/Firebag confluence was carried out. For left stream segments (segments with even numbers), values were interpolated between the station upstream of the Athabasca/Clearwater confluence and Athabasca/Firebag confluence. After obtaining TSS amounts for each segment, the average ratio of long-term means was used to estimate the concentration of silt, clay and sand in each segment. Initial concentrations for benthic segments were estimated from available bed sediment data for the Lower Athabasca River collected as part of the Regional Aquatics Monitoring Program [36].

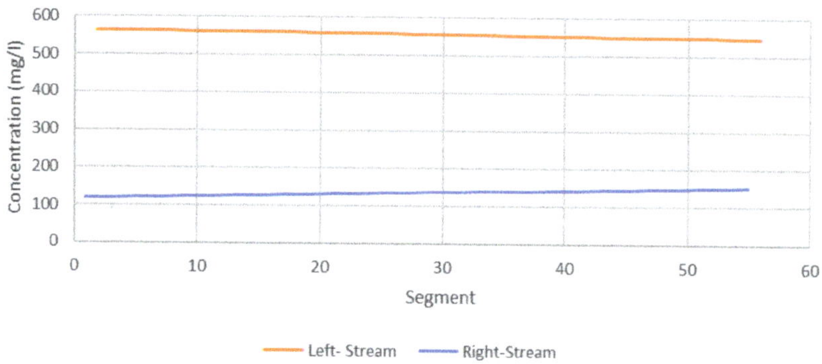

Figure 14. Initial TSS concentration along modelling domain; flow is from left to right.

The model was run until it reached an equilibrium state, after which no significant changes in the sediment concentrations in the surface water segments were observed. It was assumed that an equilibrium state has been reached when there was no significant change in sediment concentration in the water body for more than a simulation time of 24 h. For the current study, clay, silt and sand were modelled, and the results of these for both the left and right streams of the Athabasca River are shown in Figure 15.

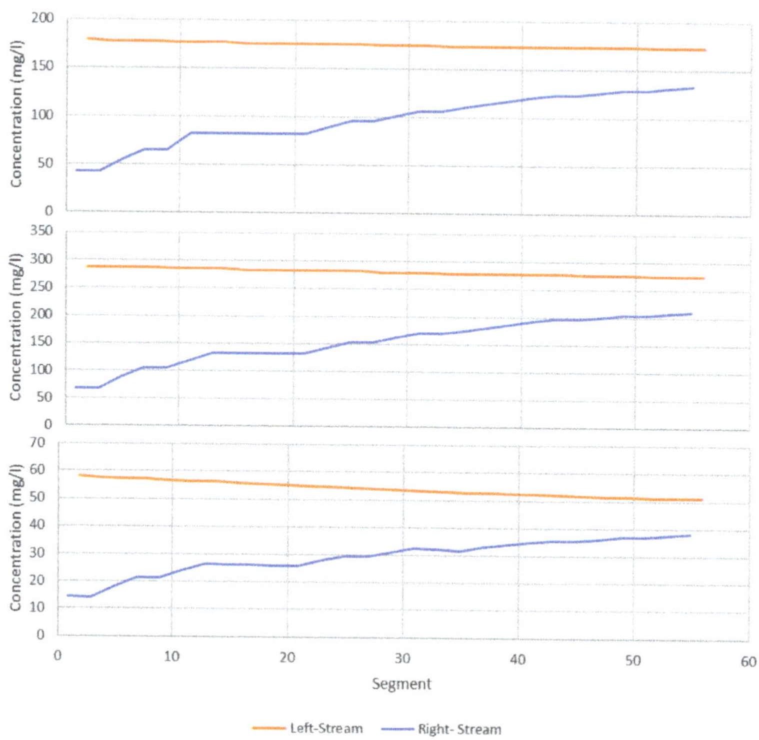

Figure 15. Simulated concentrations of clay (**top** panel), silt (**middle** panel) and sand (**bottom** panel).

The mechanics of solid transport in WASP are based on advection, and suspended sediments are only transported transversely when there is a flow routed between segments of the right and left streams. It is evident that the left stream segments have higher sediment concentrations than the right stream segments. In addition, sediment concentrations tend to drop in the left stream and increase in the right stream, both with flow direction, which is attributed to the transverse mixing, as is the case for the tracer simulations.

For all segments, including left and right streams, it was observed that, for the water columns, the amount of sediments decreases in order from silt, to clay and then sand. Due to the higher settling velocity for sand, it can be expected that there would be more sand deposits in the river compared to silt and clay. Sand is typically transported as a bedload or during higher velocity flows due to its roughness and higher density [37–39]. Due to its low settling velocity, clay is the dominant sediment texture in the water column.

At Shipyard Lake, a considerable deposition of sediment in the lake was simulated compared to the sediment in the right stream of the Athabasca River (see Figure 16). The flow through the lake is in the direction of: right stream → deep part → shallow part → right stream (further downstream). The higher sedimentation in the lake is due to the drop in flow velocity of the water entering the lake. Much of the sediment of all three textures is deposited in the deeper part of the lake with some additional deposition in the shallow lake area. The amount of solids increases in the surface benthic segments because settling dominates over resuspension in the lake.

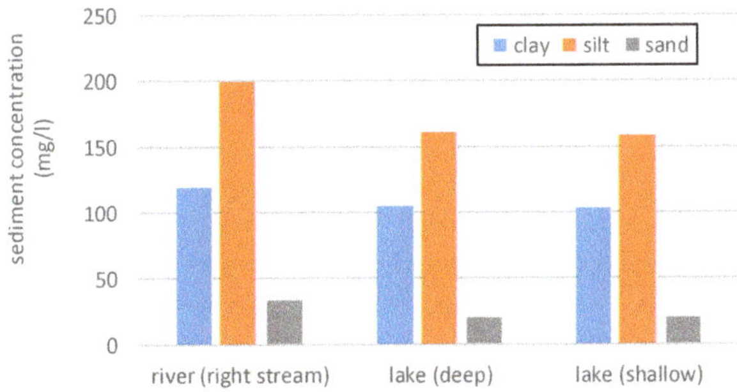

Figure 16. Water-column concentrations of clay, silt and sand in the right stream of the river and the deep and shallow portions of Shipyard Lake.

The simulations with WASP were very fast, taking only about 30 s for the model to reach steady state. This is much faster than a two-dimensional approach that was carried out in which a "relatively coarser grid was used for all of the main chemical constituent runs due to their higher computational requirements" [40]. Our approach presented here would be an alternative to creating a second grid.

8. Summary and Conclusions

A novel methodology was proposed to simulate transverse mixing along a long stream of a river network with a secondary channel and side lake. A steady-state water quality model based on two flow scenarios (600 and 2800 m^3/s) was developed to study transverse mixing and sediment deposition in the modelling domain.

The first objective of this study was to model transverse mixing in the river-lake system. Good agreement with the previously developed ARM model was observed. It was established that a quasi-2D approach is a reasonable and accurate means of modelling transverse mixing of the system with a minimal processing time compared to more complex 2D models. The approach is also well suited for scaling up to a larger river network in future work by extending the modelling domain from Fort McMurray to the Athabasca Delta.

The second objective was to quantify deposition of sediments along the system. As expected, much of the sediment load deposits are found within the secondary channel and the side lake, with less deposition in the main channel. It was observed that coarser materials tend to deposit in the deep part of the lake, where the Athabasca River water first enters the lake. Due to the longer settling time for finer sediments, they were the main sediments suspended in the main stem.

The last objective was to emphasize the importance of drawing on the output from additional models to complement the water-quality modelling exercise. To support the water-quality modelling with WASP, the hydrodynamics of the water-quality modelling domain was analyzed using HEC-RAS and ONE-D. The ONE-D model has an extended domain that incorporates recordings from gauge stations. The calibrated and validated ARM model was also useful to calibrate the transverse mixing within the WASP model. Should we scale up to extend the modelling stretch from Fort McMurray to the Athabasca Delta, additional models, in particular a hydrological model of the Athabasca River basin, should also be included. This will be essential to predict the state of the river's water quality under a changing climate scenario in the future. The impacts of an ice cover on the river's water quality should also be considered in future research.

Author Contributions: Conceptualization, T.R. and K.-E.L.; methodology, K.-E.L.; model setup, E.A. and P.S.; calibration, P.S. and K.-E.L., validation, P.S.; formal analysis, P.S. and K.-E.L.; investigation, P.S., T.R. and K.-E.L.; resources, T.R. and K.-E.L.; data curation, T.R. and K.-E.L.; writing—original draft preparation, P.S. and K.-E.L.; writing—review and editing, T.R. and K.-E.L.; supervision, T.R. and K.-E.L.; project administration, T.R. and K.-E.L.; funding acquisition, T.R. and K.-E.L. All authors have read and agreed to the published version of the manuscript.

Funding: This research was funded by Alberta Environment and Parks and the Integrated Modelling Program for Canada (IMPC) under the University of Saskatchewan's Global Water Futures (GWF) program.

Institutional Review Board Statement: Not applicable.

Informed Consent Statement: Not applicable.

Conflicts of Interest: The authors declare no conflict of interest. The funders had no role in the design of the study; in the collection, analyses, or interpretation of data; in the writing of the manuscript, or in the decision to publish the results.

Appendix A

Figures A1–A4 provide a wider range of total suspended sediment (TSS) vs. flow (Q) for Athabasca River upstream of Fort McMurray (Figure A1), Clearwater River (Figure A2), Athabasca River upstream of Firebag River (Figure A3 and Athabasca River at Old Fort (Figure A4).

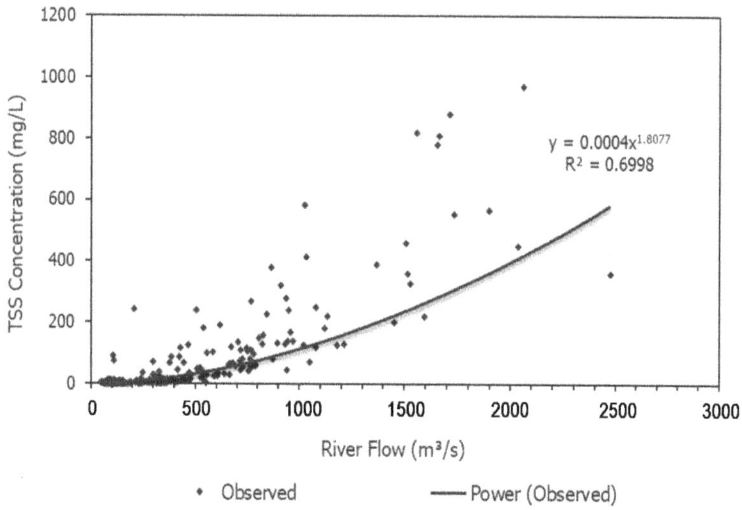

Figure A1. Total suspended sediment vs. flow for Athabasca River upstream of Fort McMurray.

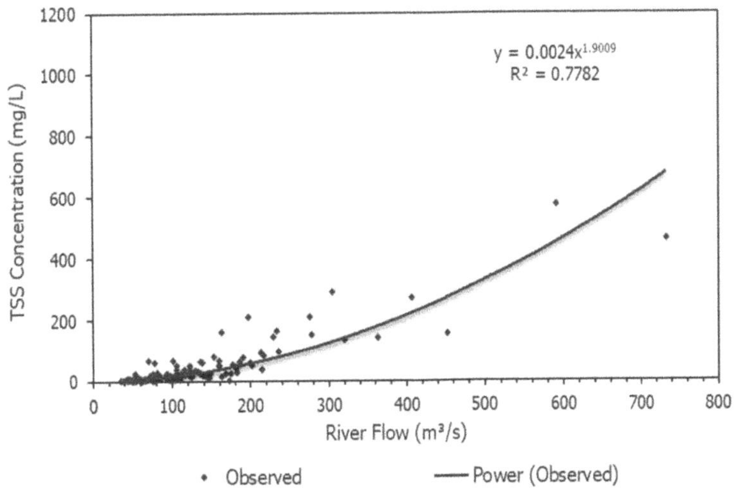

Figure A2. Total suspended sediment vs. flow for Clearwater River.

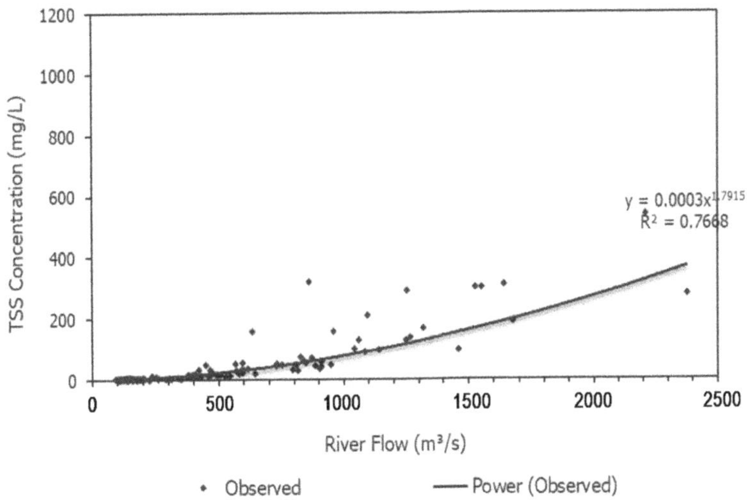

Figure A3. Total suspended sediment vs. flow for Athabasca River upstream of Firebag River.

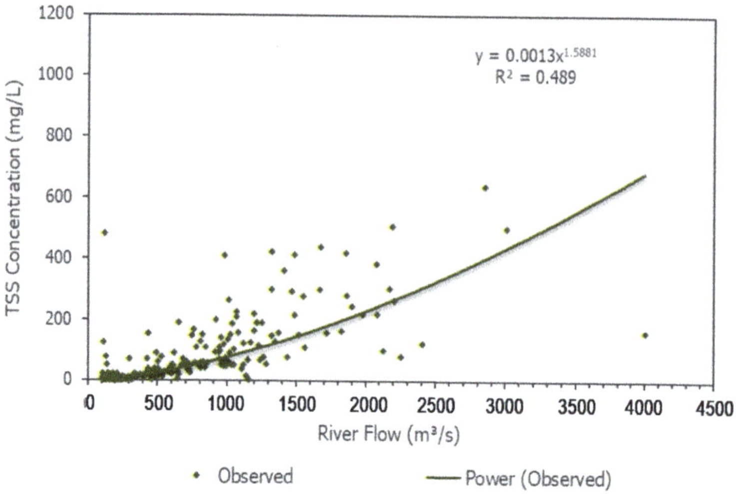

Figure A4. Total suspended sediment vs. flow for Athabasca River at Old Fort.

References

1. ECCC Canada-Alberta Joint Oil Sands Monitoring Program-Suspended Sediment Monitoring Disclaimer and Notes, Environment and Climate Change Canada. 2021. Available online: https://donnees.ec.gc.ca/data/substances/monitor/sediment-oil-sands-region/sediment-quality-mainstem-and-tributaries-oil-sands-region/?lang=en (accessed on 1 November 2021).
2. Gupta, A.; Farjad, B.; Wang, G.; Eum, H.; Dubé, M. *Integrated Environmental Modelling Framework for Cumulative Effects Assessment*; University of Calgary Press: Calgary, AB, Canada, 2021.
3. EPA. Water Quality Analysis Simulation Program (WASP), Environmental Protection Agency. Available online: https://www.epa.gov/ceam/water-quality-analysis-simulation-program-wasp (accessed on 1 November 2021).
4. Wool, T.; Ambrose, R.B., Jr.; Martin, J.L.; Comer, A. WASP 8: The Next Generation in the 50-year Evolution of USEPA's Water Quality Model. *Water* **2020**, *12*, 1398. [CrossRef] [PubMed]
5. Lindenschmidt, K.-E.; Pech, I.; Baborowski, M. Environmental risk of dissolved oxygen depletion of diverted flood waters in river polder systems-a quasi-2D flood modelling approach. *Sci. Total Environ.* **2009**, *407*, 1598–1612. [CrossRef] [PubMed]
6. Lindenschmidt, K.-E. Quasi-2D approach in modelling the transport of contaminated sediments in floodplains during river flooding-model coupling and uncertainty analysis. *Environ. Eng. Sci.* **2008**, *25*, 333–352. [CrossRef]
7. Lindenschmidt, K.-E.; Huang, S.; Baborowski, M. A quasi-2D flood modelling approach to simulate substance transport in polder systems for environmental flood risk assessment. *Sci. Total Environ.* **2008**, *397*, 86–102. [CrossRef] [PubMed]
8. Sehnert, C.; Lindenschmidt, K.-E. The effect of upstream river hydrograph characteristics on zinc deposition, dissolved oxygen depletion and phytoplankton growth in a flood detention basin system. *Quat. Int.* **2009**, *208*, 158–168. [CrossRef]
9. Huang, S.; Vorogushyn, S.; Lindenschmidt, K.-E. Quasi 2D hydrodynamic modelling of the flooded hinterland due to dyke breaching on the Elbe River. *Adv. Geosci.* **2007**, *11*, 21–29. [CrossRef]
10. Huang, S.; Rauberg, J.; Apel, H.; Disse, M.; Lindenschmidt, K.-E. The effectiveness of polder systems on peak discharge capping of floods along the middle reaches of the Elbe River in Germany. *Hydrol. Earth Syst. Sci.* **2007**, *11*, 1391–1401. [CrossRef]
11. Sehnert, C.; Huang, S.; Lindenschmidt, K.-E. Quantifying structural uncertainty due to discretisation resolution and dimensionality in a hydrodynamic polder model. *J. Hydroinformatics* **2009**, *11*, 19–30. [CrossRef]
12. Cunge, J.A. Two-dimensional modeling of floodplains. In *Unsteady Flow in Open Channels*; Mahood, K., Yevjevich, V., Eds.; Water Resources Publications: Fort Collins, CO, USA, 1975; Volume 2, pp. 705–762.
13. Aureli, A.; Maranzoni, A.; Mignosa, P.; Ziveri, C. Flood hazard mapping by means of fully-2-D and quasi-2-D numerical modeling: A case study. In *Floods, From Defence to Management*; van Alphen, J., van Beek, E., Taal, M., Eds.; 3rd International Symposium on Flood Defence: Nijmegen, The Netherlands; Taylor and Francis/Balkema: Blain, France, 2006; pp. 373–382, ISBN 0415391199.
14. Asselman, N.E.M.; Van Wijngaarden, M. Development and application of a 1-D floodplain sedimentation model for the River Rhine in the Netherlands. *J. Hydrol.* **2002**, *268*, 127–142. [CrossRef]
15. OSEM. Oil Sands Environmental Monitoring. 2021. Available online: https://www.environment.alberta.ca/apps/OSEM/ (accessed on 31 July 2021).
16. Golder. *Reach-Specific Water Quality Objectives for the Lower Athabasca River*; Golder Associates Ltd. for Cumulative Environmental Management Association: Fort McMurray, AB, Canada, 2007.
17. Golder. *Aquatic Baseline Report for the Athabasca, Steepbank and Muskeg Rivers in the Vicinity of the Steepbank and Aurora Mines (Appendices)*; Golder Associates Ltd. for Suncor Inc.: Calgary, Alberta, Canada, 1996.

18. Klohn-Crippon. *Hydrology Baseline for Project Millenium*; Report PA 289.03 submitted to Suncor Energy Inc. by Klohn-Crippon, Publisher: Vancouver, BC, Canada, 13 April 1998. Available online: chttps://era.library.ualberta.ca/items/2f3407d5-193a-4a9e-b2cb-633ffd7c8a3f/view/cc88c7c8-781a-44a5-934c-ecd4aaf5d0b4/2A-2_-_Hydrology_Baseline_for_Project_Millennium.pdf (accessed on 1 November 2021).
19. Newman, K.T. An Integrated Analysis of Sediment Geochemistry and Flood History of Floodplain Lakes in the Athabasca Region. Ph.D. Thesis, University of Saskatchewan, Saskatoon, SK, Canada, 2019. Available online: https://harvest.usask.ca/handle/10388/12436 (accessed on 1 November 2021).
20. USACE. HEC-RAS River Analysis System–User's Manual Version 6.0. US Army Corps of Engineers Hydrologic Engineering Center. 2021. Available online: https://www.hec.usace.army.mil/software/hec-ras/documentation/HEC-RAS_6.0_Users_Manual.pdf (accessed on 31 July 2021).
21. Dailey, J.E.; Harleman, D.R.F. *Numerical Model for the Prediction of Transient Water Quality in Estuary Networks*; Report # 158, RT72-72; Hydrodynamics Laboratory, Massachusetts Institute of Technology: Cambridge, MA, USA, 1972.
22. Morse, B. *St-Lawrence River Water-Levels Study: Application of the ONE-D Hydrodynamic Model*; Transport Canada, Waterways Development Div., Canadian Coast Guard: Ottawa, ON, Canada, 1990.
23. Matte, P.; Secretan, Y.; Morin, J. Hydrodynamic modeling of the St. Lawrence fluvial estuary, 1: Model setup, calibration and validation. *J. Waterw. Port Coast. Ocean Eng.* **2017**, *143*, 04017010. [CrossRef]
24. Sydor, M.; DeBoer, A.; Cheng, T. *Developing a Mathematical Flow Simulation Model for the Peace-Athabasca Delta*; Environment Canada and Alberta Environment: Edmonton, AB, Canada, 1979. Available online: http://www.worldcat.org/oclc/70366515 (accessed on 1 November 2021).
25. Sydor, M.; Kassem, A.; Hamory, T. Evaluating potential implications of the proposed Devils Lake outlets on the quality of receiving water. In Proceedings of the Canadian Water Resources Association 55th Annual Conference, Winnipeg, Manitoba, 11–14 June 2002.
26. Chowdhury, E.H.; Hassan, Q.K.; Achari, G.; Gupta, A. Use of Bathymetric and LiDAR Data in Generating Digital Elevation Model over the Lower Athabasca River Watershed in Alberta, Canada. *Water* **2017**, *9*, 19. [CrossRef]
27. Lindenschmidt, K.-E.; Sabokruhie, P.; Rosner, T. Modelling transverse mixing of sediment and vanadium in a river impacted by oil sands mining operations. *J. Hydrol. Reg. Stud.* submitted.
28. Fischer, H.B.; List, J.E.; Koh, C.R.; Imberger, J.; Brooks, N.H. *Mixing in Inland and Coastal Waters*; Academic Press Inc.: Cambridge, MA, USA, 1979.
29. Doneker, R.L.; Jirka, G.H. CORMIX User Manual: A Hydrodynamic Mixing Zone Model and Decision Support System for Pollutant Discharges into Surface Waters, EPA-823-K-07-001, December 2007. Available online: http://www.mixzon.com/downloads/ (accessed on 1 November 2021).
30. Trillium. *Measurement of Transverse Mixing Coefficients in the Lower Athabasca River*; Prepared by Trillium Engineering and Hydrographics Inc. for the Cumulative Environmental Management Association (CEMA) Surface Water Working Group: Hood River, Oregon, USA, 2003.
31. Golder. *Athabasca River Model Update and Reach Segmentation*; Prepared by Golder Associates Ltd. for Cumulative Environmental Management Association: Fort McMurray, AB, Canada, 2004.
32. Rosner, T.D. Athabasca River Model Documentation. 2021. Available online: https://fourelements.sharepoint.com/sites/AthabascaRiverModel (accessed on 1 November 2021).
33. Dimotakis, P.E. The mixing transition in turbulent flows. *J. Fluid Mech.* **2000**, *409*, 69–98. [CrossRef]
34. Sreenivasan, K.R. Turbulent mixing: A perspective. *Proc. Natl. Acad. Sci. USA* **2019**, *116*, 18175–18183. [CrossRef] [PubMed]
35. Ferguson, R.I.; Church, M.A. A simple universal equation for grain settling velocity. *J. Sediment. Res.* **2004**, *74*, 933. [CrossRef]
36. RAMP. *Regional Aquatics Monitoring Program in Support of the Joint Oil Sands Monitoring Plan*; Final 2015 Program Report; Prepared for Alberta Environmental Monitoring Evaluation and Reporting Agency by Hatfield Consultants, Kilgour and Associates Ltd., and Western Resource Solutions: Fort McMurray, AB, Canada, April 2016.
37. van Rijn, L.C. Unified View of Sediment Transport by Currents and Waves. I: Initiation of Motion, Bed Roughness, and Bed-Load Transport. *J. Hydraul. Eng.* **2007**, *133*, 649–667. [CrossRef]
38. Dey, S.; Papanicolaou, A. Sediment threshold under stream flow: A state-of-the-art review. *ASCE J. Civ. Eng.* **2008**, *12*, 45–60. [CrossRef]
39. Maggi, F. The settling velocity of mineral, biomineral, and biological particles and aggregates in water. *J. Geophys. Res. Ocean.* **2013**, *118*, 2118–2132. [CrossRef]
40. Kashyap, S.; Dibike, Y.; Shakibaeinia, A.; Prowse, T.; Droppo, I. Two-dimensional numerical modelling of sediment and chemical constituent transport within the lower reaches of the Athabasca River. *Environ. Sci. Pollut. Res.* **2017**, *24*, 2286–2303. [CrossRef] [PubMed]

water

MDPI

Commentary

Assessing and Mitigating Ice-Jam Flood Hazards and Risks: A European Perspective

Karl-Erich Lindenschmidt [1,*], Knut Alfredsen [2], Dirk Carstensen [3], Adam Choryński [4], David Gustafsson [5], Michał Halicki [6], Bernd Hentschel [7], Niina Karjalainen [8], Michael Kögel [3], Tomasz Kolerski [9], Marika Kornaś-Dynia [10], Michał Kubicki [11], Zbigniew W. Kundzewicz [4], Cornelia Lauschke [12], Albert Malinger [13], Włodzimierz Marszelewski [14], Fabian Möldner [3], Barbro Näslund-Landenmark [15], Tomasz Niedzielski [6], Antti Parjanne [16], Bogusław Pawłowski [14], Iwona Pińskwar [4], Joanna Remisz [6], Maik Renner [17], Michael Roers [17], Maksymilian Rybacki [13], Ewelina Szałkiewicz [13], Michał Szydłowski [9], Grzegorz Walusiak [6], Matylda Witek [6], Mateusz Zagata [18] and Maciej Zdralewicz [13]

1 Global Institute for Water Security, School of Environment & Sustainability, University of Saskatchewan, Saskatoon, SK S7N 3H5, Canada
2 Department of Civil and Environmental Engineering, Norwegian University of Science and Technology, 7491 Trondheim, Norway
3 Institute of Hydraulic Engineering and Water Resources Manage, Nuremberg University of Technology, 90489 Nuremberg, Germany
4 Faculty of Environmental Engineering and Mechanical Engineering, Poznan University of Life Sciences, 60-649 Poznań, Poland
5 Swedish Meteorological and Hydrological Institute, SE-601 76 Norrköping, Sweden
6 Department of Geoinformatics and Cartography, Faculty of Earth Sciences and Environmental Management, University of Wroclaw, 50-137 Wroclaw, Poland
7 Federal Waterways Engineering and Research Institute, 76187 Karlsruhe, Germany
8 Centre for Economic Development, Transport and the Environment for Lapland, Hallituskatu 3 B, 96101 Rovaniemi, Finland
9 Faculty of Civil and Environmental Engineering, Gdansk University of Technology, 80-233 Gdańsk, Poland
10 Institute of Meteorology and Water Management, National Research Institute, 01-673 Warsaw, Poland
11 Jacobs, 30-302 Kraków, Poland
12 Wasserstraßen- und Schifffahrtsamt Oder-Havel, 16225 Eberswalde, Germany
13 Flood and Drought Modelling Centre in Poznań, Institute of Meteorology and Water Management—National Research Institute, 60-594 Poznań, Poland
14 Department of Hydrology and Water Management, Nicolaus Copernicus University in Toruń, 87-100 Toruń, Poland
15 Swedish Civil Contingencies Agency, SE-651 81 Karlstad, Sweden
16 Ministry of Agriculture and Forestry, 00023 Helsinki, Finland
17 Hochwassermeldezentrale, Landesamt für Umwelt Brandenburg, 15236 Frankfurt (Oder), Germany
18 Department of Inland Waterways, Regional Water Management Board, 70-030 Szczecin, Poland
* Correspondence: karl-erich.lindenschmidt@usask.ca

Citation: Lindenschmidt, K.-E.; Alfredsen, K.; Carstensen, D.; Choryński, A.; Gustafsson, D.; Halicki, M.; Hentschel, B.; Karjalainen, N.; Kögel, M.; Kolerski, T.; et al. Assessing and Mitigating Ice-Jam Flood Hazards and Risks: A European Perspective. *Water* **2023**, *15*, 76. https://doi.org/10.3390/w15010076

Academic Editor: Olga Petrucci

Received: 30 November 2022
Revised: 14 December 2022
Accepted: 16 December 2022
Published: 26 December 2022

Abstract: The assessment and mapping of riverine flood hazards and risks is recognized by many countries as an important tool for characterizing floods and developing flood management plans. Often, however, these management plans give attention primarily to open-water floods, with ice-jam floods being mostly an afterthought once these plans have been drafted. In some Nordic regions, ice-jam floods can be more severe than open-water floods, with floodwater levels of ice-jam floods often exceeding levels of open-water floods for the same return periods. Hence, it is imperative that flooding due to river ice processes be considered in flood management plans. This also pertains to European member states who are required to submit renewed flood management plans every six years to the European governance authorities. On 19 and 20 October 2022, a workshop entitled "Assessing and mitigating ice-jam flood hazard and risk" was hosted in Poznań, Poland to explore the necessity of incorporating ice-jam flood hazard and risk assessments in the European Union's Flood Directive. The presentations given at the workshop provided a good overview of flood risk assessments in Europe and how they may change due to the climate in the future. Perspectives from Norway, Sweden, Finland, Germany, and Poland were presented. Mitigation measures, particularly the artificial breakage of river ice covers and ice-jam flood forecasting, were shared. Advances in ice processes were also presented at the workshop, including state-of-the-art developments in tracking ice-floe velocities using particle tracking velocimetry, characterizing hanging dam ice, designing new ice-control structures, detecting, and monitoring river ice covers using composite imagery from both radar and optical satellite sensors, and calculating ice-jam flood hazards using a stochastic modelling approach.

Keywords: European Union's Floods Directive; hydro-electric power; ice-jam flood hazard; ice-jam flood risk; space-borne remote sensing

---

1. Introduction

On 19 and 20 October 2022, a workshop entitled "Assessing and mitigating ice-jam flood hazard and risk" was held in Poznań, Poland, hosted by the first author with sponsorship from the Global Water Futures research program (https://gwf.usask.ca/ accessed on 14 December 2022). The workshop brought together an international team of engineers, scientists, and officials from universities, research facilities, and government agencies from Europe to review the state-of-the-art developments of and explore advances in ice-jam flood hazard and risk assessments. Government agencies from central, eastern, and northern European countries (e.g., Norway, Sweden, Finland, Germany, and Poland) are in need of new tools to assess ice-jam hazards and risks in order to propose new means of mitigating consequences of ice jamming and ice-jam flooding to communities, infrastructure, and ship navigation. These are issues that will also help research and ice-flood management of rivers in other cold-region countries (e.g., U.S.A., Canada, and Russia) affected by river ice processes and ice flooding.

The Poznań workshop was opened by a welcome from Professor Klaudia Borowiak, the Dean of the Faculty of Environmental and Mechanical Engineering at the Poznań University of Life Sciences. It ran over the course of two half days with numerous presentations, as listed in Table 1. Most of the participants are shown in the group photo in Figure 1. This workshop was a follow-up to the workshop entitled "Developing an ice-jam flood forecasting system for the Oder River" held in Wroclaw, Poland on 26 and 27 November 2018 [1]. The Poznań workshop summarized in this commentary extends the capabilities of the ice-jam flood forecasting discussed in the Wroclaw workshop by to exploring methods and requirements for the assessment and mapping of ice-jam flood hazards and risks from a European perspective. An important question posed at the workshop was: should ice-jam flood hazard and risk assessment and mapping be explicitly mentioned in the EU Floods Directive, at least for members of Nordic countries and countries with continental climates? Comments from Norway, Sweden, Finland, Germany, and Poland are provided in the section "Potential of including ice-jam flood hazard and risk in the EU Floods Directive" below. The section is preceded by "Flood risk and the European Union's Floods Directive" and "Changes to flood risk and ice-jam flood risk due to future climate" to provide background information on the EU Floods Directive and to provide some context on how ice-jam flood risk may change with the climate in the future. Presentations that introduced measures to mitigate ice-jam flood risk and technical advances in ice research to help improve ice-jam flood risk characterization are also summarized in subsequent sections below. Remarks on the workshop outcomes and an outlook for future research themes related to ice-jam floods conclude the commentary.

The paper is structured to capture the topics presented at the workshop. Section 2 provides introductory material in the topic of flood risk and how it is administered in the EU Floods Directive. Section 3 provides a perspective of future flood risk in a changing climate. Perspectives from Norway, Sweden, Finland, Germany, and Poland of including ice-jam flood risk in the EU Floods Directive are provided in Section 4. Section 5 explores different techniques used to mitigate ice-jam risks, such as artificial breakage, flood warning systems, and ice-jam flood forecasting. Current technical advances in river ice research are showcased in Section 6, with topics on particle tracking velocimetry to monitor ice-jam covers, hanging dam characterization, design of ice-control structures, processing composite radar and optical space-borne remote sensing imagery for ice characterization, implementing air-borne remote sensing tools (drones) for ice-cover monitoring, and new modelling approaches in quantifying ice-jam flood hazards. Conclusions and an outlook are found at the end of the paper.

Table 1. Oral talks presented at the workshop.

Presenter	Presentation Title
Day 1	
Klaudia Borowiak	Welcoming remarks from the Dean of Environmental and Mechanical Engineering, Poznan University of Life Sciences
Mateusz Zagata	Polish perspectives on mitigating ice-jam flood risk along the Oder River
Michael Kögel, Fabian Möldner & Dirk Carstensen	A river with ice floods - the Oder river
Iwona Pinskwar & Zbigniew W. Kundzewicz	Changes in flood risk in the Odra and Vistula river basins
Marika Kornaś-Dynia & Włodzimierz Marszelewski	Monitoring of ice phenomena on the Warta River in Poznań over a 60-year period (1961–2020)
Maik Renner & Michael Roers	Challenges for operational flood warning under ice-jam conditions at the Oder River in Brandenburg
T. Niedzielski, M. Halicki, J. Remisz, G. Walusiak & M. Witek	Applying satellite altimetry over the Odra River to issue hydrological predictions at virtual stations
Michał Kubicki	River Ice Detection on High Definition Optical and Radar Satellite Sensors
Karl-Erich Lindenschmidt	Advances in ice-jam flood forecasting, risk assessment and mitigation
Day 2	
Bogusław Pawłowski	Causes of the February 2021 ice jams in the upper Włocławek reservoir
Knut Alfredsen	Ice flood risk reduction in Norway
David Gustafsson	Ice-jam flood risk in Sweden
Tomasz Kolerski	Assessment of the ice jam severity based on the numerical models results
Adam Choryński, Iwona Pinskwar & Zbigniew Kundzewicz	Flood risk reduction in Poland
Maksymilian Rybacki	Modeling flood scenarios from ice jams using MIKE 21 Flow Model FM
Ewelina Szałkiewicz	Determination of the probability of exceedance of maximum ice-jam water states

Figure 1. Most of the participants of the workshop: 1. Cornelia Lauschke, 2. Maciej Zdralewicz, 3. Tomasz Kolerski, 4. Grzegorz Walusiak, 5. Maik Renner, 6. Joanna Remisz, 7. Michal Halicki, 8. Matylda Witek, 9. Dirk Carstensen, 10 Michal Kubicki, 11. Bogusław Pawłowski, 12. Michael Roers, 13. Karl-Erich Lindenschmidt, 14. Tomasz Niedzielski, 15. Michal Szydlowski, 16. Ewelina Szalkiewicz, 17. Wlodzimierz Marszelewski, 18. Iwona Pinskwar, 19. Maksymilian Rybacki, 20. Michael Kögel, 21. Zbigniew W. Kundzewicz, 22. Marika Kornas-Dynia, 23. Mateusz Zagata, and 24. Adam Chorynski (photo taken by Bogusław Pawłowski).

2. Flood Risk and the European Union's Floods Directive

According to the Intergovernmental Panel on Climate Change (IPCC), risk is defined as the potential for consequences where an object of value is at stake and where the outcome is uncertain. The components of such risks are hazards, exposure, and vulnerability. Referring to Figure 2, a hazard is the potential occurrence of a physical event that may cause adverse impacts. The "presence of people, livelihoods, species or ecosystems, environmental functions, services, and resources, infrastructure, or economic, social, or cultural assets in places and settings that could be adversely affected" [2] is referred to as exposure. Vulnerability can be understood as the "propensity or predisposition to be adversely affected" [2].

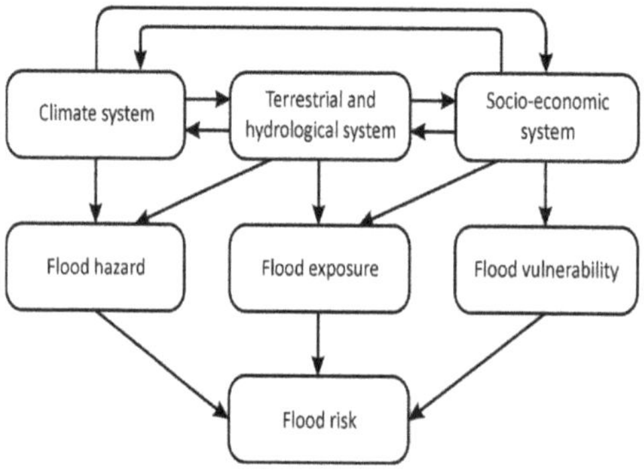

Figure 2. Components of flood risk [3].

> *"The purpose of [the European Union's Floods] Directive is to establish a framework for the assessment and management of flood risks, aiming at the reduction of the adverse consequences for human health, the environment, cultural heritage and economic activity associated with floods in the [European] Community."* [4]; Chapter 1, Article 1.

Implementation of the Floods Directive is on a six-year cycle with the European Union's member states required to follow three consecutive steps: (i) preliminary flood risk assessments, (ii) hazard maps and flood risk maps, and (iii) flood risk management plans (FRMP). The ultimate goal is to devise a FRMP for each member state to:

1. Reduce flood risk by maintaining and increasing the existing water catchment retention capacity, eliminating, or avoiding an increase in land development in areas of particular flood risk, determining the conditions for the possible development of areas protected by embankments, and avoiding growth and determining development conditions in areas with a low probability of flood occurrence.
2. Reducing the existing flood risk by limiting development in floodplains and reducing the vulnerability of facilities and communities to flood risk.
3. Improving the flood risk management systems which require implementation of forecasting and issuance of warnings about meteorological and hydrological hazards, making the responses of people, companies, and public institutions to floods more effective, increasing resilience to return to preflood states quickly, requiring effective postflood analyses, building legal and financial instruments that discourage or encourage certain behaviors to increase flood safety, and building educational programs to improve awareness and knowledge of the sources of flood hazards and risks.

3. Changes to Flood Risk and Ice-Jam Flood Risk Due to the Future Climate

Observational data show that extreme precipitation is becoming more extreme, nearly on a global scale [5]. Observed connections between heavy precipitation and air temperature broadly agree with the Clausius–Clapeyron law, foreseeing an increase in the vapor-holding capacity of the atmosphere at a rate of approximately 6–7% per 1 °C warming. This sensitivity may be much higher for precipitation on a subdaily scale, i.e., close to 14% per degree of warming for hourly precipitation [6,7].

However, this increase in extreme rainfall does not reflect higher discharges in rivers and decreases are observed at many stations. Globally, the number of stations with significant decreasing trends prevails over the number of stations with significant increasing trends [8].

In the future, along with the warming climate, atmospheric water vapor content is likely to increase, hence the potential for the occurrence of heavy precipitation is on the rise. According to Huo et al. [9], short-duration extreme precipitation may prevail over long-duration extremes. Extreme precipitation events with intensity exceeding the infiltration capacity and the conveyance capacity of the system will very likely result in urban and flash floods of increasing frequency and magnitude. Strong increases in the frequencies of extreme precipitation events (from the 95th to the 99.97th percentile) based on an analysis of observations was presented by Myhre et al. [10]. The total precipitation from these intense events almost doubles per degree of warming, mainly due to changes in frequency. As shown by Hettiarachchi et al. [11] such extremely intense short-duration events will cause flooding in most areas.

The frequency and magnitude of fluvial (river) floods is expected to increase in many regions, but the statement that these kinds of floods are on the rise has not been substantiated. However, projections for the future indicate a greater increase in land areas where river floods become more frequent, compared to the fraction of areas for which fluvial floods will decrease.

At this time there is a lack of consistency between the trends observed in river discharges, which do not indicate an increase, and the model-based projections for the future, which show increasing trends. However, climate change has accelerated, and some changes may yet reveal themselves, so the change expected for the far future could be considerably different from the trend that is now observed [3].

4. Potential of Including Ice-Jam Flood Hazards and Risks in the EU Floods Directive

4.1. Norwegian Perspective

Although ice jams are mentioned in the new guidelines for flood zone maps currently, after hearing from the Norwegian Water Resources and Energy Directorate (NVE), it appears that the guidelines may not be very clear on a specific procedure for mapping ice-induced flooding. For regulated rivers, there are some restrictions on operations to avoid ice-induced floods, particularly the placement and adjustment of intakes to prevent exposing open water to frigid air and avoid huge frazil generation. Operational restrictions are mainly related to freeze-up flooding due to frazil and anchor ice, but also to the risk of breakup in the case of accidental shutdowns and water being released into bypass reaches. In the proposed guidelines, it is stated that ice jams rarely exceed the 200-year open water flood in Norway, and therefore it is not a central component in the flood zone mapping procedure, since the 200-year level is critical in the Norwegian building code. The guidelines acknowledge that it can be an issue and recommend considering ice where problems with ice have been observed in the past. This can particularly be an issue where water is diverted from the river (but not on the 200-year return period level). At best, ice jams that have threatened hydro-power generation or are caused by hydropower shutdowns are archived using maps or local images, an example of an ice-jam event that occurred on the Svorkmo River in Norway is shown in Figure 3.

4.2. Swedish Perspective

The Swedish Meteorological and Hydrological Institute (SMHI) also sees the main issue of ice-jam flood hazards to be frazil ice and ice-jam impacts on hydropower (occasionally there are also problems in nonregulated rivers), but it has no official mandate or task to provide ice information. However, the institute does produce ice-breakup forecasts for the Torne River, including forecasts of breakup dates and a "severity degree" factor indicating the risk for ice-jam complications. The forecasts are provided in collaboration and conjunction with forecasts provided by the Finnish Environment Institute (SYKE), which acquires the necessary ice depth data. Both institutes are involved in research projects to further develop river ice forecasting and monitoring capabilities. An example of a flood hazard map indicating some past flood extents due to ice jams is given in Figure 4.

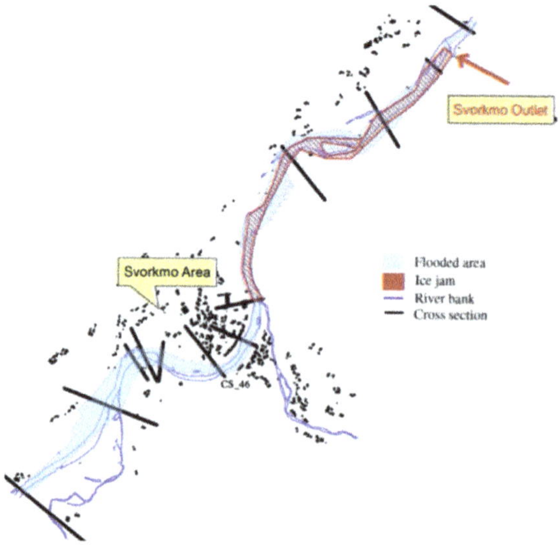

Figure 3. Ice jam on the Svormo River affecting the operation of a downstream hydro-power generation facility (from [12]).

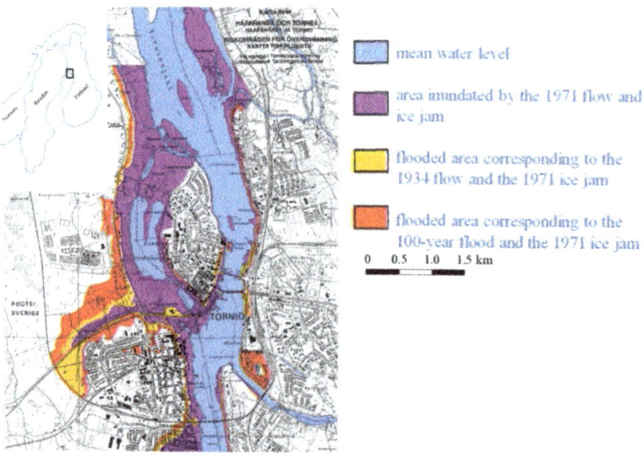

Figure 4. Extent of past ice-jam flood events [13].

4.3. Finnish Perspective

The EU Floods Directive does consider floods caused by frazil ice and ice jams, even though they are not specifically mentioned in the directive itself. However, according to the directive, the preliminary flood risk assessment should be based on information regarding ice-jam or frazil-ice floods that have occurred in the past and provide an outlook for potential future floods. Hence, from a Finnish perspective, the Floods Directive does take ice-jam floods sufficiently into account.

In Finland, there are areas of potentially significant flood risk, where designation is based on ice-jam flood risk (risk for significant adverse consequences by ice jams causing floods). Additionally, goals and measures in the flood risk management plans for those areas are targeted specifically for preventing ice jamming or taking action when there are rapidly rising floodwaters due to ice jams. Finnish government authorities have also prepared flood hazard and risk maps for those areas with past ice-jam flood occurrences (roughly based on similar open water flow velocities), for example for Tornio in Figure 5.

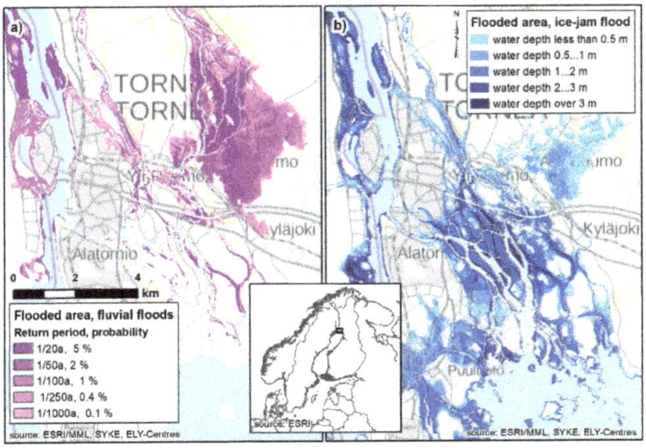

Figure 5. Flood extent for (**a**) different fluvial flood probabilities and (**b**) a severe ice-jam flood (flow = 3100 m^3/s and +2 m sea water level) at the Torne River outlet. (source: SYKE).

4.4. German Perspective

The Oder River is the river that is most often affected by ice jams in Germany. This area, in the most eastern part of the country, experiences continental temperatures in winter. Ice jams generally occur during freeze-up events and are greatly influenced by backwater from the wind setup in the Baltic Sea entering the river's mouth at Szczecin. Ice blockage, in particular, is problematic for flooding at river structures such as bridge piers and weirs. Efforts to release ice jams are carried out by German and Polish ice breakers, but the flow of released ice is hampered by the very mild slope of the lower reaches of the Oder River. Vulnerabilities exist at dikes which have breached in the past due to ice-jam flood events. "The stretch along the Oder Bruch, formally an inland delta drained for agricultural use, is particularly vulnerable due to its containment through dikes and the sediment accretion of the riverbed to elevations higher than the surrounding land. A catastrophic event of extended flooding throughout the adjacent low-lying area of the Oder Bruch occurred in March 1947, in which ice jams caused backwaters to overtop and breach dikes along the Oder Bruch at two locations, with breach widths of over 100 m. Flooding was extensive, leading to the evacuation of 20,000 people" [14]. The fact that ice jamming has become less frequent along the Oder River in recent decades, plus the advances in flood protection and ice defense measures, warning systems, and corresponding disaster control measures, have led to a lack of perception by the people of the dangers and risks of ice-jam floods along the German Oder riverbanks.

4.5. Polish Perspective

At the preliminary flood risk assessment stage and based on the Floods Directive Reporting Guidance 2018 to the European Commission, Poland can identify different types of floods e.g., fluvial and pluvial (from rivers or overland runoff), sea water (flooding of land by water from the sea, estuaries, or coastal lakes) and artificial water-bearing infrastructure (flooding of land by water arising from artificial, water-bearing infrastructure, or failure of such infrastructure). Floods resulting from blockages or restrictions may also be identified, a category which would include ice-jam floods. Other mechanisms that fall into this category include blockages of sewerage systems, restrictive channel structures such as bridges or culverts, and natural occurrences, such as landslides.

It is still uncertain if flood hazard and risk maps for ice jams will be developed for Poland in the future. Much depends on the results of the next preliminary flood risk assessment and the decisions of government authorities. Currently, the Institute of Meteorology and Water Management, a National Research Institute, is working on various aspects of flood protection, including mathematical modeling of ice jams and determining flood hazards from ice phenomena. Figure 6 shows the results of a preliminary study modelling flood hazard areas from ice jams along the test section of the Oder River.

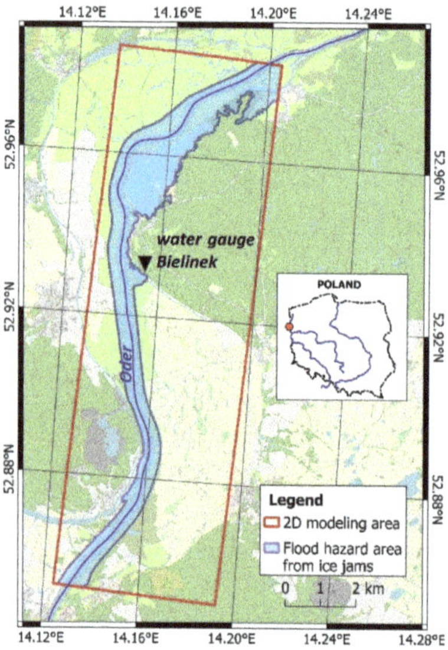

Figure 6. Flooded area along the Oder River at Bielinek, Poland (base map from OpenStreetMap https://www.openstreetmap.org/#map=7/52.012/16.414 accessed on 14 December 2022).

In the upcoming publishing of the river basin management plan for Poland, locations of past ice jams have been identified and mapped for the Vistula River basin, shown in Figure 7. Concentrations of ice-jam locations are indicative of river stretches with a higher propensity for ice-jam flood hazards.

In regard to the changing climate in Poland, in the headwaters of the Vistula River, in the Carpathian Mountains in southern Poland, mean annual air temperatures at the Beskid Zywiecki station have increased by more than 2 °C over the past 40 years. Generally, increases in annual air temperature for stations in the upper Vistula River basin were at the rate of +0.13 °C per 10 years to +0.29 °C per 10 years (based on the period 1951–2015) [15]. Annual total precipitation has also increased, with an increasing trend of approximately 100 mm over the past 50 years. Lupikasza et al. [15] found trends in annual precipitation at ten stations located in the upper Vistula River basin to vary from −7.2 mm per 10 years up to 16.5 mm per 10 years for the period 1951–2015. The intensity of precipitation has also changed, with the number of days of precipitation totaling more than 5 and 10 mm/day increasing over the same time period. Pinskwar et al. [16] also found that, for the area of the upper Vistula River basin, the number of days with precipitation equal to or above 10 mm as well as 20 mm increased in the period 1991–2015 in comparison to 1961–1990. This has repercussions on the flows along the Vistula River and the degree of substances transported from the catchment area into the receiving waters. The more intense rainfalls lead to a greater supply of eroded material to the rivers, exacerbated by the increased weathering of rocks and erosion due to rising air temperatures. The additional sediment transported in rivers can lead to increased accretion of the riverbed, particularly at the inlet of reservoirs, as is the case for the Wloclawek Reservoir showcased below in the Section "Ice characterization of a hanging dam".

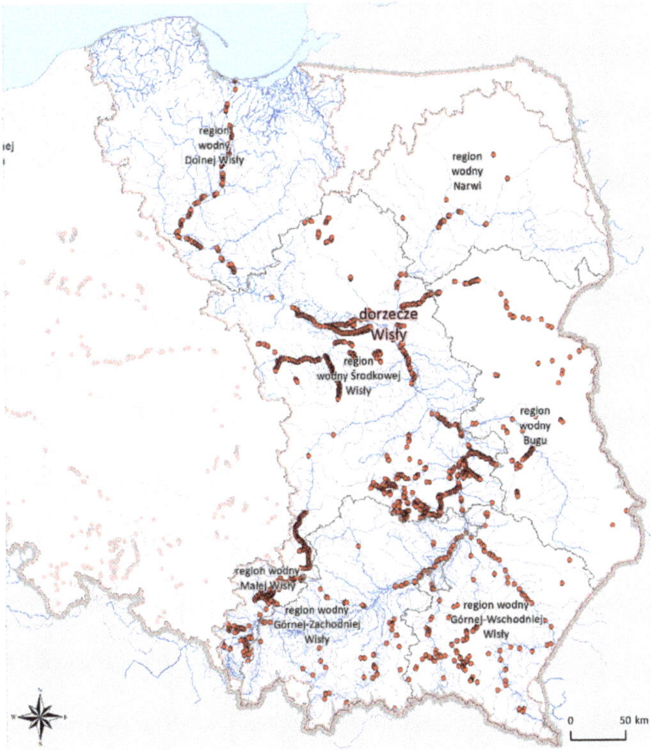

Figure 7. Possible locations of past ice-jam events along the Vistula River and its tributaries (second update of the river basin management plans).

Ice phenomena and phenology have also changed in rivers in Poland. Since 1960, ice phenomena generally appear on the Warta River in December or January of each winter, and the trend is that their occurrences have been delayed by approximately three days per decade at the gauge at Poznań. The ice season generally ends in February or March, with the trend in the end dates occurring approximately four days earlier each decade. This leads to a progressive shortening of the ice season, as shown in Figure 8. The figure also reveals a significant increase in the trend in mean air temperatures measured in Poznań. An interesting correlation between air temperature and the duration of the ice season can be drawn between the two, as indicated in Figure 9. A suggestion was made at the workshop to model water temperature and apply a correction to the ice phenology due to urban heat islands. One of the workshop participants mentioned that most gauges in Poland, with long-term records of ice phenology and thicknesses, are situated in urban centres, which may result in steeper trends toward shorter ice durations and thinner ice nation-wide compared to potential trends due to climatic conditions alone. The additional heat may stem from effluents such as those from wastewater treatment plants or altered air temperatures. In particular air temperature changes should be tested since a large urban area may be required for a significant effect to occur on river ice. A hydrological modelling system with the capability of simulating river temperatures, e.g., the MESH-RBM modelling system [17,18], could help determine biases in ice phenology and thicknesses when comparing "actual" water temperatures (due to climate change and urban heating) to "natural" water temperatures (due to climate change alone). Increased transport of dissolved substances, particularly from the application of fertilizers in the surrounding agricultural region, can also lead to a shortening of the ice season.

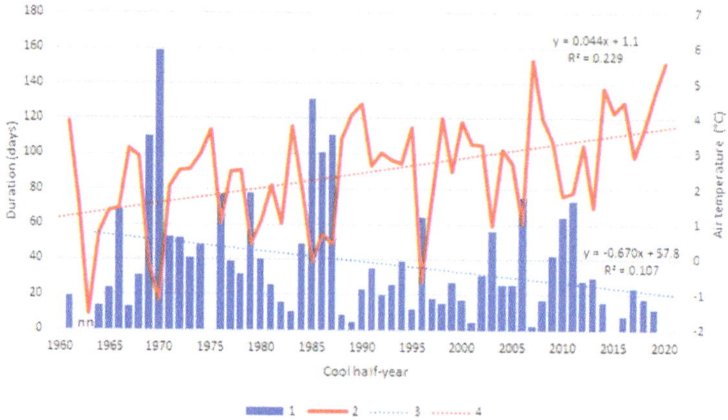

Figure 8. Changes in the duration of ice phenomena on the Warta River in Poznań against the backdrop of the average air temperature in the cool half-year (November to April) in 1961–2020; 1—ice phenomena, 2—average air temperature in the cool half-year, 3—linear trend of ice phenomena in 1964–2020 and 4—linear trend of the average air temperature in the cool half-year in 1961–2020; n—no data (source: data from IMWM-NRI and RWMB in Poznań). (data source: IMWM-NRI and RWMB in Poznań).

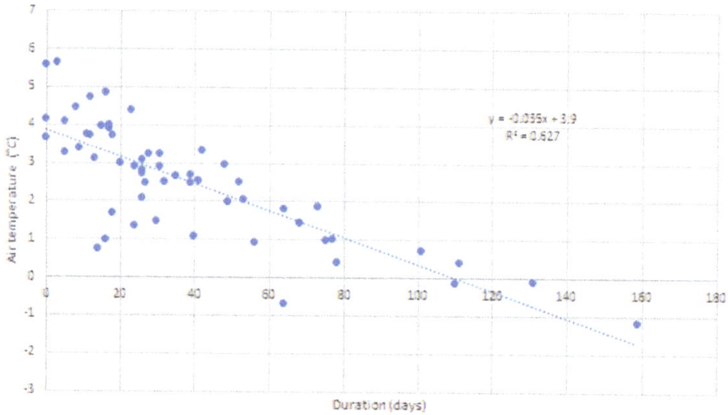

Figure 9. Correlation between average air temperature in the cool half-year (November to April) and duration of the ice phenomena on the Warta River in Poznań for the time period 1964–2020.

On the lower Vistula River, the duration of ice phenomena during the winters in the period 1960–2014 has also decreased [19]. The strongest negative trend was observed in the cross-section of the station situated immediately downstream of the river dam in Wloclawek, approximately −1.5 days/year. Negative trends of −1.64 to −1.97 days/year were also observed at other gauging stations. Negative trends were greater downstream of the Wloclawek Dam than upstream.

5. Ice-Jam Flood Risk Mitigation Measures

5.1. Artificial Ice-Cover Breakage

In order to reduce flood hazards and risks due to ice jamming, ice breakers operate along major waterways to artificially break up ice covers (see Figure 10). The icebreaking operation on the Oder River along the Polish–German border is carried out jointly by the Polish and German waterways administrations. The technical management of the breakage operation of both icebreaker fleets (seven Polish and six German with two reserve icebreakers, one Polish and one German) is exercised by the Polish administration.

Figure 10. Icebreakers releasing an ice jam on the Oder River (source: RZGW, Szczecin).

Generally, a permanent ice cover develops first on Dabie Lake in Szczecin (see map in Figure 11), where frazil ice travelling down the Oder River accumulates and juxtapositions upstream along the Oder River's main stem and its tributaries, the Warta and Lusatian Neisse rivers. Icebreaking begins with crushing the permanent ice cover on Dabie Lake and freeing a gutter through the ice cover to make room for ice floes broken upstream along the river. Frontal icebreakers are directed upriver to break the consolidated ice cover along the river, while linear icebreakers crisscross Dabie Lake to prevent the broken ice from stagnating and refreezing.

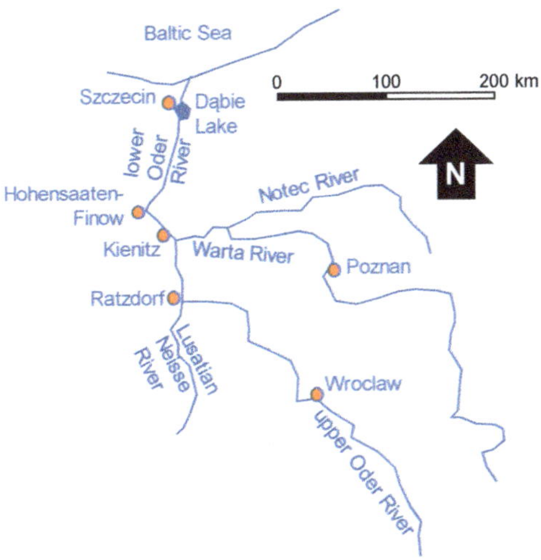

Figure 11. Main stem of the Oder River and its major tributaries (drawn by the first author).

Icebreaking on the Oder River continues upstream towards the mouth of the Warta River, where the resources are split, with the larger part of the icebreaker fleet continuing breakage along the Oder River towards the Lusatian Neisse River mouth and the remaining icebreaker force working its way up the Warta River to the Notec River mouth. Caution must be taken not to begin breakage operations too early so as not to create a large amount of ice floes flowing from the upper sections of the Oder River and its tributaries to the lower reach of the Oder River to create ice jams and, thus, risk inducing a flood artificially.

5.2. Flood Warning under Ice Conditions

Operational flood warning relies on river gauge monitoring. The federal state of Brandenburg, Germany defines four different flood alert levels, which are specified for representative river gauges used in flood reporting services, see Figure 12. An alert level is proclaimed when water levels exceed a certain alert stage and local authorities must take action. The alert levels require increasing operational flood defense actions with increasing water stages [20]: alert level AI—water level reporting service (German: *Meldebeginn*), AII—control service at flood defense infrastructure such as dikes (German: *Kontrolldienst*), AIII—guard duty (German: *Wachdienst*), and AIV—civil protection (German: *Hochwasserabwehr*). These four alert levels allow quick assessment of the potential severity of a flood across different rivers in Brandenburg. The gauges used for the alert level system require high reliability and redundancy of sensors and communication networks.

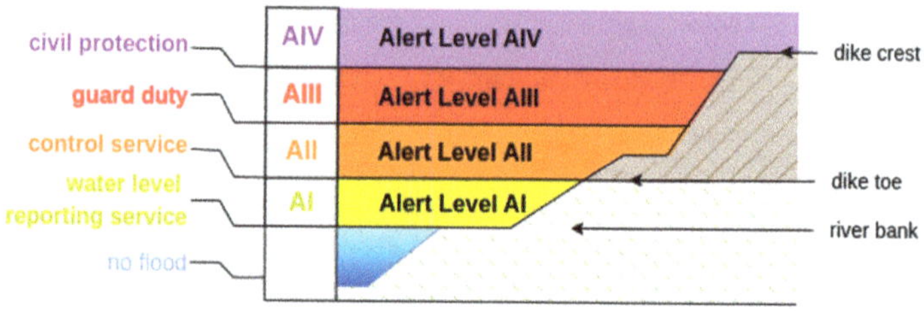

Figure 12. Alert levels for flood warnings (source: Landesamt für Umwelt Brandenburg).

Flood alert gauges are linked to specific sections of a river which are defined according to the local flood risk and hydraulic conditions, as well as the appropriate administrative units. Figure 13 highlights the river sections used for flood alerts at the lower Oder River in Brandenburg, with panel A showing the sections under normal flood conditions. Since ice-jam flood dynamics are completely different at the lower Oder River, with ice-jams moving upstream, two important adjustments have been implemented. First, river sections upstream of the gauges are used for warning (Figure 13B) and lower alert stages are defined, since ice-jams typically lead to damaged dikes and abrupt rises in water levels are expected. With these adjustments the flood warning system is more representative of ice-flood conditions.

5.3. Ice-Jam Flood Forecasting

Ice-jam flood forecasting is a key component in any flood management plan to reduce flood hazards and risks. Advances have been made in the development of ice-jam flood forecasting methodologies and systems, particularly for the Athabasca River at Fort McMurray, Alberta [21], the upper reaches of the Saint John River, New Brunswick [22], and the Sanhuhekou bend of the Yellow River in China [23]. These methodologies and systems have also been implemented successfully in ice-jam flood forecasting systems for operational use by the government of Newfoundland and Labrador for the lower (Atlantic) Churchill River [24,25] and by the government of Manitoba for the lower Red River in Manitoba [26]. Requirements for an operational ice-jam flood forecasting system for the Oder River have been laid out in [1] and the need to include such methodologies for ice conditions can be seen in Figure 14, which shows a rapid rise in the backwater levels at Hohensaaten-Finow (see map in Figure 11 for the location) caused by an ice jam downstream of that gauge in February 2021. Forecasts on the rising limb of the event grossly underestimated the water levels attained by the ice jamming since no river ice processes are integrated in the current hydraulic model used for operational forecasting. The roughness coefficients and the rating curves implemented in the model also require updating to reflect ice-jam backwater effects. An ice-jam hydraulic model has been set up for the Oder River [27] between Ratzdorf and Kienitz (see Figure 11 for locations) and needs to be extended to Dabie Lake to include ice-jam backwater effects at Hohensaaten-Finow.

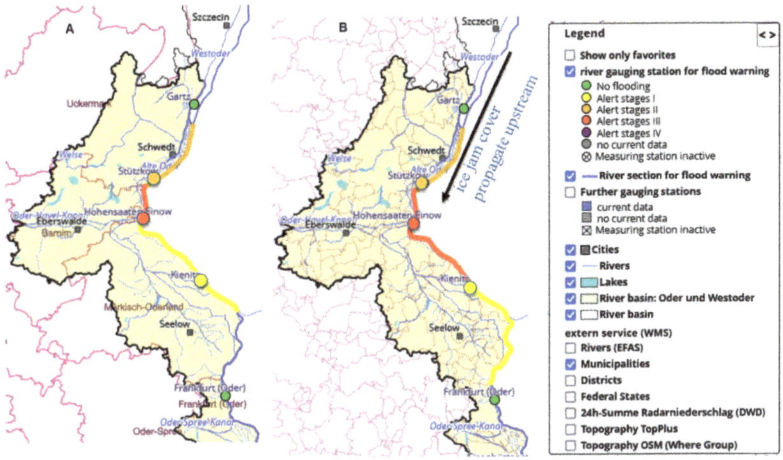

Figure 13. Tracking an ice-jam cover progression using the alert level system, which links gauge readings to upstream river reaches (source: Landesamt für Umwelt Brandenburg) with (**A**) for normal flood conditions and (**B**) for ice-jam conditions.

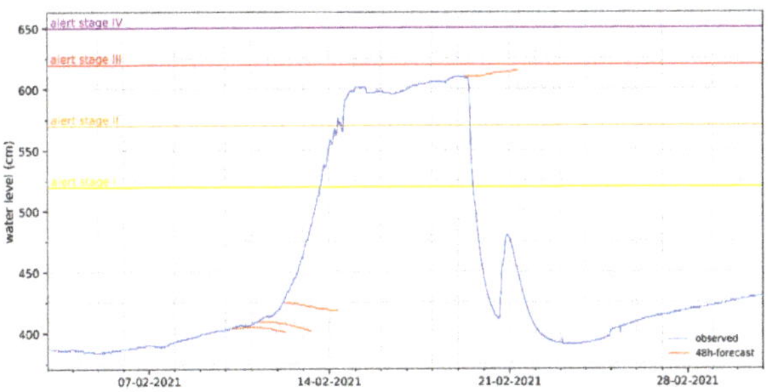

Figure 14. Measurement and forecast of water levels at the gauge at Hohensaaten-Finow, February 2021 (source: Landesamt für Umwelt Brandenburg).

6. Technical Advances in River Ice Research

6.1. Particle Tracking Velocimetry

A novelty presented at the workshop is determining flow velocities of ice using particle tracking velocimetry, which allows the velocities and trajectories of ice floes to be measured remotely. The method tracks the flow velocities of many ice floes using a sequence of images. It includes measuring the camera position and orientation (camera pose), automatic extraction of the water area for feature searching, particle detection and filtering, particle tracking and filtering, and scaling the tracks of (ice floe) velocities [28]. Figure 15 shows such trajectories with velocities of ice floes across the west channel of the lower Oder River during the February 2021 ice-jam event.

Figure 15. Trajectory and flow velocities of ice floes from the ice-jam event of February 2021 in the west channel of the lower Oder River (source: Technische Hochschule Nürnberg).

6.2. Ice Characterization of a Hanging Dam

In February 2021, a severe ice jam occurred in the upstream end of the Wloclawek Reservoir on the Vistula River near Plock [29]. This area is one of the most ice-jam prone river stretches in Poland. Evidence points to sediment accretion at the reservoir inlet as the reason for the increase in this area's propensity to ice jamming, indicated by a shift in the rating curve over time between the years 2009 and 2020. Efforts to break up the ice jam with ice breakers were hindered by the shallow depth of the reservoir inlet. Dredging works have been cut back in recent years even though the intensity of sedimentation has increased due to greater transport of sediment from the upstream catchment area, with an accretion rate of approximately 5 cm/year downstream of Plock. Despite the shortening of the duration of the ice season, the ice-jam flood risk of this reservoir section remains high.

To determine the volume of ice in the hanging dam that caused the jamming and the thickness of the hanging dam in relation to the water depth, cross-sections of the ice with depth were surveyed at the upstream end of the hanging dam using a sounding device (weight) to penetrate the hanging dam ice. During these frazil slush penetration tests, changes in the compactness of the ice deposits constituting the hanging dam were also recorded to determine the amount of ice grounded on the reservoir bottom during the ice-jam event. Three grades of compactness were classified:

- firm accumulations—sounder must be driven into the ice by force.
- compact accumulations—sounder remains stationary within the slush.
- loose accumulations—sounder penetrates the accumulation driven by its own weight.

The cross-section of the depths of the hanging dam is shown in Figure 16, indicating a decrease in the ice compactness with depth. A water layer was still evident between the bottom of the hanging dam and the reservoir bottom (no grounding); however, the hanging dam did fill a substantial percentage of the cross-sectional flow area. Flow velocity on the right side of the cross-section would have been greater where loose slush ice did not deposit, whereas on the left side, flow velocities would have been less, allowing frazil ice to be deposited on the underside of the hanging dam.

6.3. Design of New Ice-Control Structure

Ice jams are initiated when ice transport conveyance is reduced locally along a river stretch, particularly in meanders or in areas where the river narrows and obstacles are present in the river (e.g., islands, bridge piers, and sand bars) [30]. A continuous high inflow of ice from upstream can also help in the initiation process of ice jams. High volumes of inflowing ice can lead to increases in ice thicknesses constricting the cross-sectional flow area, resulting in the impediment of discharge under the ice jam with an increase in water surface elevations in the section upstream of the jam. One means of reducing the influx of ice in an ice-jam prone area is to arrest the flow of ice upstream of a potential ice-jam location using an ice-control structure (ICS), shown in Figure 17 [31]. The structure mostly impedes the ice transport further downstream but not the flow of the water, which is allowed to bypass the ice accumulation. A transverse set of piers only partially spans across the channel from one bank; the piers then extend longitudinally upstream parallel to the bank to form a side channel between the longitudinal set of piers and the bank. This side channel provides a passage of water to flow around the accumulation of ice which is held back by the piers. With this design, additional space in an adjacent floodplain to bypass water around the ice accumulation is not required.

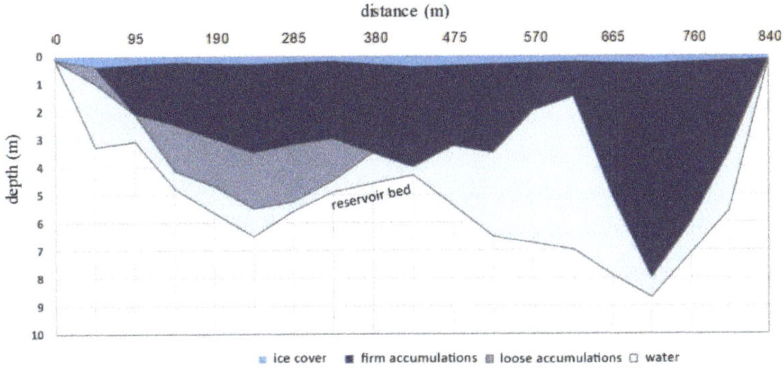

Figure 16. Cross-section of the hanging dam formed at the Wloclawek Reservoir inlet near Plock in February 2021 [29].

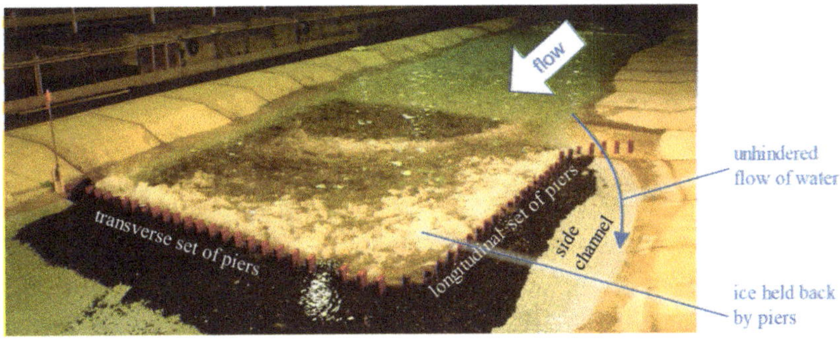

Figure 17. Ice-control structure (photo courtesy of US Army Cold Regions Research and Engineering Laboratory, Hanover, New Hampshire; used with permission).

6.4. River Ice Detection Using Optical and Radar Satellite Imagery in Tandem

The aim of river ice monitoring using satellite data is to (i) provide spatially constant, frequent information about the presence of ice along a river course, (ii) provide imagery support for water management services, and (iii) detect possible threats and natural disasters provoked by ice jams. The two satellite sensors used in this study were:

- Sentinel-1, which is a radar sensor using C-band frequency in two polarization modes (VV/VH). Two satellites (A and B), launched in 2014 and 2015, provided ~2-day revisit times across all of Europe with a spatial resolution of ~10 m and a swath width of ~250 km for the GRD product in IW mode
- Sentinel-2, which is an optical sensor with 13 spectral bands. Two satellites (A and B), launched in 2015 and 2017, provided ~5-day revisit times across Europe to acquire imagery with a spatial resolution of 10, 20, and 60 m with a swath width of ~290 km. Imagery was delivered in UTM grid cuts with overlap.

Figure 18 provides a combined product of the Sentinel-1 and Sentinel-2 images of the west and east channels of the lower Oder River. Sentinel-1 images were calibrated, speckle filtered, terrain corrected, and rescaled to dB, whereas the Sentinel-2 images underwent atmospheric correction, resampling to consistent spatial resolution of 20 m, and spectral indices calculations to strengthen the classification of the desired coverage classes (water, snow or ice, vegetation, and bare soil) before a composite of the two images was created. There are some limitations with both sensors. For example, misclassifications may occur between smooth black ice covers and open water. Misclassifications are also possible when predefined waterbeds are changed due to changing water levels. For Sentinel-2 imagery, misclassifications may also occur when differentiating between smooth black ice and open water. Image areas can also be misclassified as turbid waters or waters having an algal bloom. Clouds will also hamper Sentinel-2 image clarity.

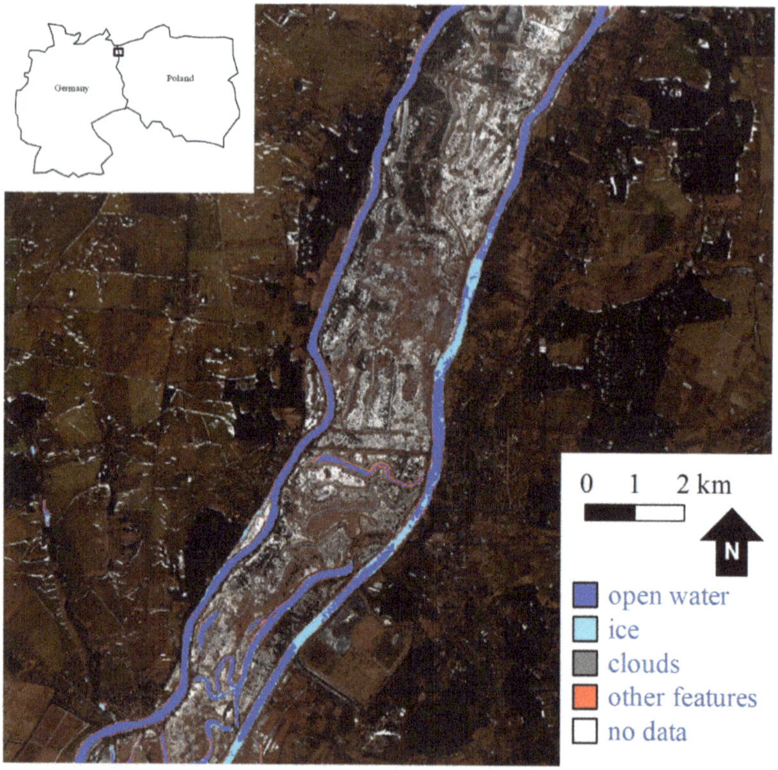

Figure 18. The combined product from Sentinel-1 and Sentinel-2 based classifications. Images credit: https://land.copernicus.eu/pan-european/biophysical-parameters/high-resolution-snow-and-ice-monitoring; images acquired on 16 February 2017.

The high-resolution sensors from the Sentinel family proved very useful in detecting ice coverage, including ice jams. The Sentinel-2 optical sensor offers better thematic accuracy and should be treated as a primary source for ice classification, whereas the Sentinel-1 radar sensor offers the possibility of providing observations even under cloudy conditions and can be used as a secondary or auxiliary source for ice classification. Ice jams are well presented in the images from both sensors. Success in these composite images have led to the creation of the high-resolution snow and ice monitoring service, developed between 2019 and 2021 under the auspices of the Europe Environmental Agency

6.5. Stochastic Model to Assess Ice-Jam Flood Hazards

The chaotic nature of ice-jam formation and flooding can be captured using a stochastic modelling approach. In this approach, a deterministic river ice hydraulics model runs many times, with each run having a different set of input values for the parameters and boundary conditions. These values are randomly selected from frequency distributions of each parameter and boundary condition. This results in an ensemble of possible ice-jamming outcomes along a river reach of interest. Such an approach has been newly developed to quantify ice-jam flood hazards and risks [32,33]. Figure 19 conceptualizes the approach, which requires frequency distributions of the boundary conditions (shown at the top of Figure 19) and parameter values (not shown) to be input to the deterministic river ice hydraulic model RIVICE (see [34,35]) for model descriptions). The boundary conditions include:

- upstream water flow Q (Figure 19a) represented by an extreme-value distribution of the flows at instantaneous water level maxima during ice-jam events,
- volume of inflowing ice accumulating in the ice jam Vice (Figure 19b), which is a function of Q, with the scatter represented by a confidence band within which random variables are selected,
- downstream water level W (Figure 19c), which is a function of the upstream discharge.
- Location of the ice-jam lodgment x (Figure 19d), which is represented by a stepped uniform distribution to capture the predisposition of ice jamming in some stretches over others.

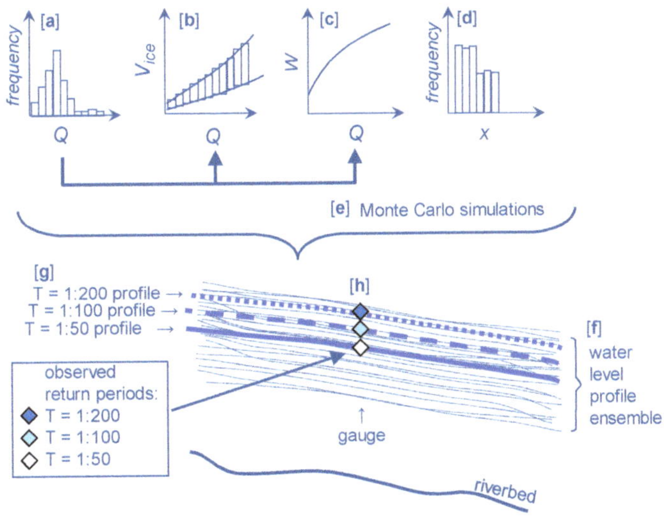

Figure 19. Conceptualization of the stochastic modelling framework for ice-jam flood hazard assessment; explanations for each subfigure are provided in the main text (drawn by the first author).

Parameters are generally uniform distributions between minimum and maximum values determined through calibration.

Using a Monte Carlo approach (Figure 19e), RIVICE runs hundreds of times, with each simulation having a different set of boundary conditions and parameter values chosen randomly from all distributions. One output is an ensemble of backwater profiles (Figure 19f), the results of which can be compiled within a probabilistic context using percentile profiles of exceedance probabilities (Figure 19g). The water level elevations at a gauge can then be compared to the annual exceedance probabilities of the levels recorded at the gauge (Figure 19h). Discrepancies between the simulated and recorded exceedance probabilities can be reduced by adjusting the percentage of the confidence band in the Q vs.

Vice relationship with the processes, including the Monte Carlo simulations, having to be repeated.

Referring to Figure 20, artificially breaking up the ice cover, for example using an ice breaker as described in the section "Artificial ice-cover breakage" above, can be implemented as an option to mitigate ice-jam flood hazards and risks. This scenario can be simulated within the stochastic modelling framework by removing those stretches in the lodgment location distribution (Figure 20D). It is assumed that ice cannot lodge to form an ice jam in areas that have been artificially broken. Rerunning the Monte Carlo analysis should lead to a change in the elevations of the percentile profiles of the backwater level ensemble.

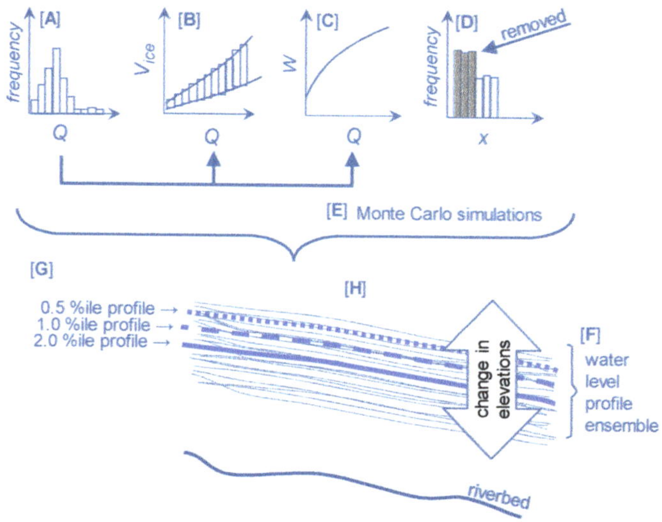

Figure 20. Adjustment in the boundary condition frequency distribution within the stochastic modelling framework when considering mitigation options, such as artificially breaking the ice cover to hinder ice-jam lodgments; explanations for each subfigure are provided in the main text (drawn by the first author).

At the workshop, research was also presented on the testing of different solutions applied to probability analyses of ice jams and ice covers. Time series of the maximum water stages were determined using: (i) extreme-value approaches using annual maximum water staging [36], and (ii) the peak-over-threshold (POT) method. The main idea of the POT method is to prepare a series of maxima on the basis of all events occurring in the analyzed time period that exceed an assumed threshold value, the so-called cut-off level [37]. The application of the POT method makes it possible to include, within the time series, the fact that ice phenomena may occur several times in one year, and, within the empirical probability, the fact that ice phenomena do not occur every year [38]. The final probability of exceedance for the POT time series was determined using the following tested distributions: log-normal, Gumbel, Pearson III, Gamma, log-Gamma, and Pareto.

6.6. Monitoring with UAVs

Growing access to unmanned aerial vehicles (UAVs), also known as drones, opens new possibilities to observe ice on rivers and reservoirs. Even low-cost UAVs are now capable of taking nadir photographs with predefined frontal and side overlap. Such images are standard input data for the structure-from-motion (SfM) algorithm, the products of which are dense point clouds and the resulting digital surface models as well as orthophotomaps. Recently, this popular approach has been adopted to determine the spatial extent of snow [39] or ice [40] as well as to reconstruct snow depth [41] or ice thickness [40].

Figure 21 presents aerial imagery of two frozen reservoirs in the Izerskie Mountains (southwest Poland) as well as two fragments of the unfrozen Oder River (west Poland). It is apparent from the figure that the visual analysis of imagery leads to the differentiation between frozen (top row in Figure 21) and unfrozen water (bottom row in Figure 21), enabling the detection of the presence of ice. Additionally, it is simple to discriminate between spatially uneven ice cover on water reservoirs in Rozdroże Izerskie and Polana Izerska and snow-covered banks or bare land. The knowledge about snow depth and extent and ice thickness and extent, acquired just after collecting UAV data within the concept of rapid mapping [42], may be useful, for instance, to assess the risks of avalanches, snow-melt floods, or ice jams.

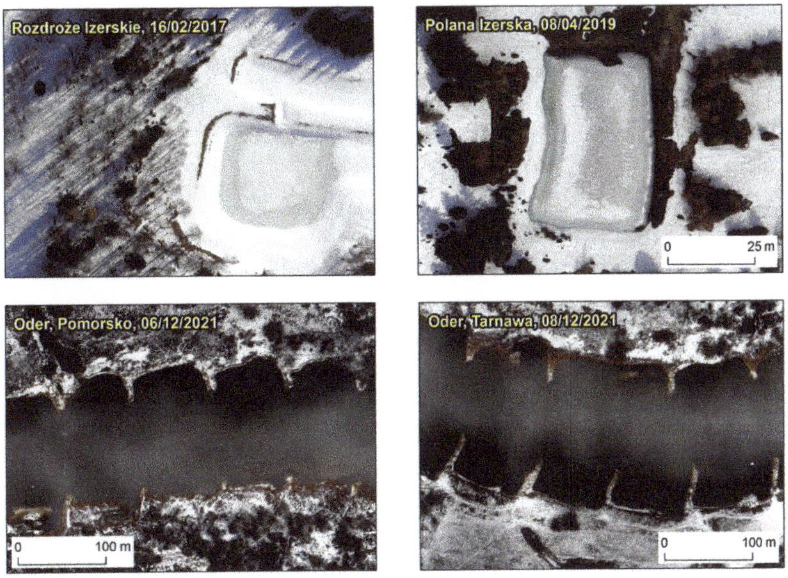

Figure 21. Fragment of single aerial image taken in central projection by a UAV over Rozdroże Izerskie in southwest Poland showing two interconnected frozen reservoirs (**top left**), fragment of the SfM-based orthophotomap of Polana Izerska in southwest Poland centered on a frozen reservoir (**top right**), fragments of the SfM-based orthophotomaps of the Oder River in west Poland in Pomorsko (**bottom left**) and Tarnawa (**bottom right**) showing an unfrozen river channel (source: Department of Geoinformatics and Cartography, University of Wrocław).

7. Conclusions

The workshop brought together many scientists and government officials who were involved in the field of river ice in their work and research. The venue provided an opportunity to present ideas and exchange knowledge in the field of ice-jam flood hazards and risks and how the subject was approached and applied in each of the EU member's countries. One key takeaway message from the workshop was that ice-jam floods are important components in the flood hazard and risk assessment and should be catalogued in the Flood Management Plans of the EU Floods Directive deliverables, but ice-jam floods do not need to be explicitly expressed within the directive itself. This may be partially due to the fact that, in rivers of northern and eastern countries, which are members of the European Union, the floodwater levels for a certain annual exceedance probability (or return periods) from ice jamming is generally lower than for those of open-water floods. A more comprehensive examination of other rivers is required for this statement to be thoroughly conclusive. Depending on the region, ice-jam flood hazards and risks have different foci, for example, hydropower operations in Norway and shipping navigation in Poland and Germany.

At the workshop, many new advances were also presented on monitoring and mitigating ice-jam flood hazards and risks, including application with particle tracking velocimetry, hanging dam characterization, ice-control structure design, remote sensing of ice covers, and modelling ice-jam flood hazards and hazard reductions. As an outlook, areas that need research furtherance include:

- safely measuring flows under covers of loose ice accumulations,
- incorporating near-ground (trail cameras), aerial (drones), and space-borne (satellites) remote sensing imagery into integrated monitoring systems for quick response to ice-jam flood hazard developments, and
- real-time monitoring of ice-cover elevations as a proxy for ice-thickness measurements.

8. Outlook

The workshop focused more on the technical aspects of flood risk assessment. A follow-up workshop could include the social aspects of ice-jam flood risk [43], for example community resilience [44] and socioeconomic vulnerabilities [45]. Research has also been carried out with agent-based modelling (ABM) to incorporate both technical and social aspects in flood risk assessments and management, on the individual [46], household [47] and regional [48] levels, in policy and decision making. An application of ABM specific to ice-jam flood risk assessment and mitigation is currently being explored by Ghoreishi et al. [49,50].

Author Contributions: Conceptualization, K.-E.L., K.A., D.C., A.C., D.G., M.H., B.H., N.K., M.K. (Michael Kögel), T.K., M.K.-D., M.K. (Michał Kubicki), Z.W.K., C.L., A.M., W.M., F.M., B.N.-L., T.N., A.P., B.P., I.P., J.R., M.R. (Maik Renner), M.R. (Michael Roers), M.R. (Maksymilian Rybacki), E.S., M.S., G.W., M.W., M.Z. (Mateusz Zagata) and M.Z. (Maciej Zdralewicz); investigation, K.-E.L., K.A., D.C., A.C., D.G., M.H., B.H., N.K., M.K. (Michael Kögel), T.K., M.K.-D., M.K. (Michał Kubicki), Z.W.K., C.L., A.M., W.M., F.M., B.N.-L., T.N., A.P., B.P., I.P., J.R., M.R. (Maik Renner), M.R. (Michael Roers), M.R. (Maksymilian Rybacki), E.S., M.S., G.W., M.W., M.Z. (Mateusz Zagata) and M.Z. (Maciej Zdralewicz); writing—original draft preparation, K.-E.L.; writing—review and editing, K.-E.L., K.A., D.C., A.C., D.G., M.H., B.H., N.K., M.K. (Michael Kögel), T.K., M.K.-D., M.K. (Michał Kubicki), Z.W.K., C.L., A.M., W.M., F.M., B.N.-L., T.N., A.P., B.P., I.P., J.R., M.R. (Maik Renner), M.R. (Michael Roers), M.R. (Maksymilian Rybacki), E.S., M.S., G.W., M.W., M.Z. (Mateusz Zagata) and M.Z. (Maciej Zdralewicz); visualization, K.-E.L., K.A., D.C., A.C., D.G., M.H., M.K. (Michael Kögel), T.K., M.K.-D., M.K. (Michał Kubicki), Z.W.K., W.M., F.M., T.N., B.P., I.P., J.R., M.R. (Maik Renner), M.R. (Michael Roers), M.R. (Maksymilian Rybacki), E.S., G.W., M.W., and M.Z. (Maciej Zdralewicz); supervision, K.-E.L., D.C. and Z.W.K.; project administration, K.-E.L., D.C. and Z.W.K.; funding acquisition, K.-E.L. All authors have read and agreed to the published version of the manuscript.

Funding: We would like to thank John Pomeroy, director of the Global Water Futures program at the University of Saskatchewan, for his sponsorship to partly fund this workshop. The drone-related fragments of the manuscript are supported by the National Science Centre of Poland through the research project no. 2020/38/E/ST10/00295—carried out by Tomasz Niedzielski, Matylda Witek, Joanna Remisz, Michał Halicki, and Grzegorz Walusiak—within the Sonata BIS programme.

Institutional Review Board Statement: Not applicable.

Informed Consent Statement: Not applicable.

Data Availability Statement: Not applicable.

Acknowledgments: We would like to extend a special thanks to all the participants for attending and contributing to the workshop.

Conflicts of Interest: The authors declare no conflict of interest.

References

1. Lindenschmidt, K.-E.; Carstensen, D.; Fröhlich, W.; Hentschel, B.; Iwicki, S.; Kögel, M.; Kubicki, M.; Kundzewicz, Z.W.; Lauschke, C.; Łazarów, A.; et al. Development of an Ice Jam Flood Forecasting System for the Lower Oder River—Requirements for Real-Time Predictions of Water, Ice and Sediment Transport. *Water* **2019**, *11*, 95. [CrossRef]
2. Agard, J.; Schipper, E.L.F. Editorial Board Co-Chairs, 2014. Glossary Annex II 1757. In *IPCC 2014: Climate Change 2014: Impacts, Adaptation, and Vulnerability. Part A: Global and Sectoral Aspects. Contribution of Working Group II to the Fifth Assessment Report of the Intergovernmental Panel on Climate Change*; Field, C.B., Barros, V.R., Dokken, D.J., Mach, K.J., Mastrandrea, M.D., Bilir, T.E., Chatterjee, M., Ebi, K.L., Estrada, Y.O., Genova, R.C., et al., Eds.; Cambridge University Press: Cambridge, UK; New York, NY, USA, 2014.
3. Kundzewicz, Z.W.; Pińskwar, I.; Brakenridge, G. Changes in river flood hazard in Europe: A review. *Hydrol. Res.* **2017**, *49*, 294–302. [CrossRef]
4. EU Directive 2007/60/EC of the European Parliament and the Council of 23 October 2007 on the Assessment and Management of Flood Risks, 2007. Available online: https://eur-lex.europa.eu/legal-content/EN/TXT/HTML/?uri=CELEX:32007L0060&from=EN (accessed on 20 November 2022).
5. Sun, Q.; Zhang, X.; Zwiers, F.; Westra, S.; Alexander, L.V. A Global, Continental, and Regional Analysis of Changes in Extreme Precipitation. *J. Clim.* **2021**, *34*, 243–258. [CrossRef]
6. Lenderink, G.; Van Meijgaard, E. Increase in hourly precipitation extremes beyond expectations from temperature changes. *Nat. Geosci.* **2008**, *1*, 511–514. [CrossRef]
7. Berg, P.M.V.D.; Moseley, C.; O Haerter, J. Strong increase in convective precipitation in response to higher temperatures. *Nat. Geosci.* **2013**, *6*, 181–185. [CrossRef]
8. Kundzewicz, Z.W.; Pińskwar, I. Are Pluvial and Fluvial Floods on the Rise? *Water* **2022**, *14*, 2612. [CrossRef]
9. Huo, R.; Li, L.; Chen, H.; Xu, C.-Y.; Chen, J.; Guo, S. Extreme Precipitation Changes in Europe from the Last Millennium to the End of the Twenty-First Century. *J. Clim.* **2021**, *34*, 567–588. [CrossRef]
10. Myhre, G.; Alterskjær, K.; Stjern, C.W.; Hodnebrog, Ø.; Marelle, L.; Samset, B.H.; Sillmann, J.; Schaller, N.; Fischer, E.; Schulz, M.; et al. Frequency of extreme precipitation increases extensively with event rareness under global warming. *Sci. Rep.* **2019**, *9*, 16063. [CrossRef]
11. Hettiarachchi, S.; Wasko, C.; Sharma, A. Increase in flood risk resulting from climate change in a developed urban watershed – the role of storm temporal patterns. *Hydrol. Earth Syst. Sci.* **2018**, *22*, 2041–2056. [CrossRef]
12. Timalsina, N.; Beckers, F.; Alfredsen, K. Modelling winter operational strategies of a hydropower system. *Cold Reg. Sci. Technol.* **2016**, *122*, 1–9. [CrossRef]
13. Zachrisson, G. Torne River—Proposals for Damage Prevention Measures (In Swedish). SMHI in Collaboration With the Finnish Water and Environment Agency, 1989. Available online: https://www.smhi.se/polopoly_fs/1.163738!/Hydrologi_25%20Sv%C3%A5ra%20islossningar%20i%20Torne%C3%A4lven.%20F%C3%B6rslag%20till%20skadef%C3%B6rebyggande%20%C3%A5tg%C3%A4rder.pdf (accessed on 14 December 2022).
14. Lindenschmidt, K.-E.; Huokuna, M.; Burrell, B.C.; Beltaos, S. Lessons learned from past ice-jam floods concerning the challenges of flood mapping. *Int. J. River Basin Manag.* **2017**, *16*, 457–468. [CrossRef]
15. Łupikasza, E.; Niedźwiedź, T.; Pinskwar, I.; Ruiz-Villanueva, V.; Kundzewicz, Z.W. Observed Changes in Air Temperature and Precipitation and Relationship between them, in the Upper Vistula Basin. In *Flood Risk in the Upper Vistula Basin*; Springer: Berlin/Heidelberg, Germany, 2016; pp. 155–187. [CrossRef]
16. Pińskwar, I.; Choryński, A.; Graczyk, D.; Kundzewicz, Z.W. Observed changes in extreme precipitation in Poland: 1991–2015 versus 1961–1990. *Theor. Appl. Clim.* **2018**, *135*, 773–787. [CrossRef]
17. Morales-Marín, L.; Sanyal, P.; Kadowaki, H.; Li, Z.; Rokaya, P.; Lindenschmidt, K. A hydrological and water temperature modelling framework to simulate the timing of river freeze-up and ice-cover breakup in large-scale catchments. *Environ. Model. Softw.* **2019**, *114*, 49–63. [CrossRef]
18. Morales-Marín, L.; Rokaya, P.; Sanyal, P.; Sereda, J.; Lindenschmidt, K. Changes in streamflow and water temperature affect fish habitat in the Athabasca River basin in the context of climate change. *Ecol. Model.* **2019**, *407*, 108718. [CrossRef]
19. Pawlowski, B. *Course of Ice Phenomena on the Lower Vistula River in 1960–2014*; Scientific Publishing House of the Nicolaus Copernicus University in Torun: Torun, Poland, 2017; p. 176. (In Polish)
20. HWMDV Ordinance on the Establishment of a Warning and Alarm Service to Protect Against Water Hazards and for the Transmission of Flood Reports) (German = Verordnung über Die Errichtung Eines Warn—Und Alarmdienstes Zum Schutz Vor Wassergefahren Und Zur Übermittlung Von Hochwassermeldungen—Hochwassermeldedienstverordnung [HWMDV]), 9 September 1997. Available online: https://bravors.brandenburg.de/verordnungen/hwmdv_2016 (accessed on 14 November 2022).
21. Lindenschmidt, K.-E.; Rokaya, P.; Das, A.; Li, Z.; Richard, D. A novel stochastic modelling approach for operational real-time ice-jam flood forecasting. *J. Hydrol.* **2019**, *575*, 381–394. [CrossRef]
22. Das, A.; Budhathoki, S.; Lindenschmidt, K.-E. A stochastic modelling approach to forecast real-time ice jam flood severity along the transborder (New Brunswick/Maine) Saint John River of North America. *Stoch. Environ. Res. Risk Assess.* **2022**, *36*, 1903–1915. [CrossRef]

23. Liu, W.; Lü, H.; Lindenschmidt, K.-E.; Xü, K.; Zhu, Y.; He, C.; Wang, X.; Xie, B. Hazard assessment and prediction of ice-jam flooding for a river regulated by reservoirs using an integrated probabilistic modelling approach for a river regulated by reservoirs using an integrated probabilistic modelling approach. *J. Hydrol.* **2022**, *615*, 128611. [CrossRef]
24. Lindenschmidt, K.-E.; Brown, D.; Khan, A.A.; Khan, H.; Khayer, M.; McArdle, S.; Mostofi, S.; Naumov, A.; Pham, T.; Weiss, A. A novel fully-operational fully-automated real-time ice-jam flood forecasting system. In Proceedings of the 25th IAHR International Symposium on Ice, Trondheim, Norway, 23–25 November 2020. Available online: https://www.iahr.org/library/infor?pid=8570 (accessed on 20 November 2022).
25. Lindenschmidt, K.-E.; Khan, A.A.; Khan, H.; Khayer, M. Parallelisation of a river ice hydraulic model to improve performance of an ice-jam flood forecasting system for the lower Churchill River in Labrador. In Proceedings of the 26th IAHR International Symposium on Ice, Montréal, QU, Canada, 19–23 June 2022.
26. Williams, B.S.; Das, A.; Johnston, P.; Luo, B.; Lindenschmidt, K.-E. Measuring the skill of an operational ice jam flood forecasting system. *Int. J. Disaster Risk Reduct.* **2021**, *52*, 102001. [CrossRef]
27. Kögel, M.; Das, A.; Marszelewski, W.; Carstensen, D.; Lindenschmidt, K.-E. Feasibility study for forecasting ice jams along the Oder River. *Wasserwirtschaft* **2017**, *5*, 20–28. (In German). Available online: https://www.springerprofessional.de/machbarkeitsstudie-zur-vorhersage-von-eisstau-auf-der-oder/12243598 (accessed on 15 November 2022). [CrossRef]
28. Eltner, A.; Sardemann, H.; Grundmann, J. Technical Note: Flow velocity and discharge measurement in rivers using terrestrial and unmanned-aerial-vehicle imagery. *Hydrol. Earth Syst. Sci.* **2020**, *24*, 1429–1445. [CrossRef]
29. Pawlowski, B.; Kolerski, T. Zator na Wisle w rejonie Plocka w lutym 2021 r. *Gospod. Wodna* **2022**, 6–10. Available online: https://mostwiedzy.pl/pl/publication/zator-na-wisle-w-rejonie-plocka-w-lutym-2021-r,157992-1 (accessed on 15 October 2022). [CrossRef] [PubMed]
30. Shen, H.T.; Gao, L.; Kolerski, T.; Liu, L. Dynamics of ice jam formation and release. *J. Coast. Res.* **2008**, *10052*, 25–32. [CrossRef]
31. Tuthill, A.; Ashton, G.; Hendershot, P.; Quadrini, J. Grasse River Ice Control Structure, Physical Model Study. Presented at the 19th IAHR International Symposium on Ice, Vancouver, BC, Canada, 6–11 July 2008. Available online: https://inis.iaea.org/search/search.aspx?orig_q=RN:40000053 (accessed on 17 September 2022).
32. Lindenschmidt, K.-E. *River Ice Processes and Ice Flood Forecasting—A Guide for Practitioners and Students*, 2nd ed. Springer Nature Switzerland AG: Berlin/Heidelberg, Germany, 2024; *in preparation*.
33. Lindenschmidt, K.-E. Extension and refinement of a stochastic modelling approach to assess ice-jam flood hazard. *Hydrol. Res. accepted*.
34. Lindenschmidt, K.-E. RIVICE—A non-proprietary, open-source, one-dimensional river-ice and water-quality model. *Water* **2017**, *9*, 314. [CrossRef]
35. Lindenschmidt, K.-E. *River Ice Processes and Ice Flood Forecasting—A Guide for Practitioners and Students*; Springer Nature Switzerland AG: Berlin/Heidelberg, Germany, 2020; p. 267. [CrossRef]
36. Stanley, S.J.; Gerard, R. Probability analysis of historical ice jam flood data for complex reach: A case study. *Can. J. Civ. Egineering* **1992**, *19*, 875–885. [CrossRef]
37. Bacová-Mitková, V.; Onderka, M. Analysis of extreme hydrological events on the Danube using the peak over threshold method. *J. Hydrol. Hydromech.* **2010**, *58*, 88–101. [CrossRef]
38. Byczkowski, A.; Banasik, K.; Hejduk, L. Obliczanie przeplywów powodziowych o okreslonym prawdopo-dobienstwie przekroczenia. *Infrastrukt. I Ekol. Teren. Wiej.* **2008**, *5*, 199–208.
39. Niedzielski, T.; Spallek, W.; Witek-Kasprzak, M. Automated Snow Extent Mapping Based on Orthophoto Images from Unmanned Aerial Vehicles. *Pure Appl. Geophys.* **2018**, *175*, 3285–3302. Available online: https://link.springer.com/article/10.1007/s00024-018-1843-8 (accessed on 17 October 2022). [CrossRef]
40. Alfredsen, K.; Haas, C.; Tuhtan, J.A.; Zinke, P. Brief Communication: Mapping River ice using drones and structure from motion. *Cryosphere* **2018**, *12*, 627–633. [CrossRef]
41. Mizinski, B.; Niedzielski, T. Fully-automated estimation of snow depth in near real time with the use of unmanned aerial vehicles without utilizing ground control points. *Cold Reg. Sci. Technol.* **2017**, *138*, 63–72. [CrossRef]
42. Tampubolon, W.; Reinhardt, W. UAV data processing for rapid mapping activities. *Int. Arch. Photogramm. Remote Sens. Spat. Inf. Sci.* **2015**, *40*, 371–377. Available online: https://www.int-arch-photogramm-remote-sens-spatial-inf-sci.net/XL-3-W3/371/2015/ (accessed on 10 October 2022). [CrossRef]
43. Aznar-Crespo, P.; Aledo, A.; Melgarejo-Moreno, J.; Vallejos-Romero, A. Adapting Social Impact Assessment to Flood Risk Management. *Sustainability* **2021**, *13*, 3410. [CrossRef]
44. Nofal, O.M.; van de Lindt, J.W. Understanding flood risk in the context of community resilience modeling for the built environment: Research needs and trends. *Sustain. Resilient Infrastruct.* **2022**, *7*, 171–187. [CrossRef]
45. Chakraborty, L.; Thistlethwaite, J.; Minano, A.; Henstra, D.; Scott, D. Leveraging Hazard, Exposure, and Social Vulnerability Data to Assess Flood Risk to Indigenous Communities in Canada. *Int. J. Disaster. Risk Sci.* **2021**, *12*, 821–838. [CrossRef]
46. Erdlenbruch, K.; Bonté, B. Simulating the dynamics of individual adaptation to floods. *Environ. Sci. Policy* **2018**, *84*, 134–148. [CrossRef]
47. Haer, T.; Botzen, W.J.W.; de Moel, H.; Aerts, J.C.J.H. Integrating Household Risk Mitigation Behavior in Flood. Risk Analysis: An Agent-Based Model Approach. *Risk Anal.* **2017**, *37*, 1977–1992. [CrossRef]

48. de Ruig, L.T.; Haer, T.; de Moel, H.; Brody, S.D.; Botzen, W.J.W.; Czajkowski, J.; Aerts, J.C.J.H. How the USA can benefit from risk-based premiums combined with flood protection. *Nat. Clim. Chang.* **2022**, *12*, 995–998. [CrossRef]
49. Ghoreishi, M.; Das, A.; Lindenschmidt, K.-E. Advancing ice-jam flood risk management: Integrating dynamic adaptive behavior using an agent-based modelling approach.
50. Ghoreishi, M.; Das, A.; Lindenschmidt, K.-E. Advancing Ice-jam Flood Risk Assessment Using Agent-based Modeling: A Conceptual Framework on Human Adaptation. AGU Fall Meeting Abstracts, 2022. Available online: https://agu.confex.com/agu/fm22/meetingapp.cgi/Paper/1164013 (accessed on 10 October 2022).

MDPI
St. Alban-Anlage 66
4052 Basel
Switzerland
Tel. +41 61 683 77 34
Fax +41 61 302 89 18
www.mdpi.com

Water Editorial Office
E-mail: water@mdpi.com
www.mdpi.com/journal/water